Lecture Notes in Engineering

Edited by C. A. Brebbia and S. A. Orszag

8

Linda M. Abriola

Multiphase Migration of Organic Compounds in a Porous Medium

A Mathematical Model

Springer-Verlag Berlin Heidelberg GmbH

Series Editors
C. A. Brebbia · S. A. Orszag

Consulting Editors
J. Argyris · K.-J. Bathe · A. S. Connor · J. Connor · R. McCrory
C. S. Desai · K.-P. Holz · F. A. Leckie · L. G. Pinder · A. R. S. Pont
J. H. Seinfeld · P. Silvester · P. Spanos · W. Wunderlich · S. Yip

Author
Linda M. Abriola
Department of Civil Engineering
University of Michigan
Ann Arbor, Michigan 48109-2125
USA

ISBN 978-3-540-13694-1 ISBN 978-3-642-82343-5 (eBook)
DOI 10.1007/978-3-642-82343-5

Library of Congress Cataloging in Publication Data

Abriola, Linda M.
Multiphase migration of organic compounds in a porous medium.
(Lecture notes in engineering ; 8)
1. Water, Underground – Pollution – Mathematical models.
2. Organic water pollutants – Mathematical models.
3. Porous materials – Mathematical models.
I. Title. II. Series.
TD426.A27 1984 628.1'68 84-13970

Lecture Notes in Engineering

The Springer-Verlag Lecture Notes provide rapid (approximately six months), refereed publication of topical items, longer than ordinary journal articles but shorter and less formal than most monographs and textbooks. They are published in an attractive yet economical format; authors or editors provide manuscripts typed to specifications, ready for photo-reproduction.

To my mother and father

Abstract

In recent years, great attention has focused on the contamination of the subsurface by organic chemicals. This problem is geographically widespread and persistent. A literature review of contamination case histories and present modeling techniques reveals the need for a more comprehensive approach to the modeling of the chemical contamination process. This approach should be capable of tracing the *multiphase* migration of a pollutant (i.e. its migration as a solute, a gas, and a non-aqueous phase). In this thesis, such an approach is developed and implemented in the construction of a numerical simulator.

Four separate phases are included in the development: solid (soil), water, gas, and contaminant. The contaminant phase may be composed of two distinct components - one volatile and the other non-volatile. Transfer of the volatile component to the water and/or gas phases is permitted. As a starting point in the analysis, the microscopic mass balance law of continuum mechanics is averaged over a representative elementary volume to produce a macroscopic mass balance equation for each system component. Based on the physical characteristics of these components, various constitutive relations and approximations can be introduced. Incorporation of these relations into the balance laws yields a system of three nonlinear partial differential equations.

This system of equations is not amenable to solution by analytical means. Approximate solutions to these equations, however, can be sought at specific points by replacing the differential operators by finite difference operators. The system of equations is thereby reduced to a system of implicit nonlinear algebraic equations in discreet unknowns. A Newton-Raphson iteration scheme provides an effective technique for the solution of these equations. Development of a one-dimensional computer model proceeds along these lines.

In order to apply this finite difference model to a specific problem, a number of equation parameters must be evaluated. These parameters include three-phase relative permeabilities, saturations, partition coefficients, and mixture densities and viscosities. Once expressions for these parameters are obtained, the numerical model may be used to simulate various one-dimensional contamination scenarios. Pollution of an unsaturated soil column by a petroleum mixture and migration of TCE in a water-saturated column are considered. Convergence and mass balance properties of the scheme are examined for each of these problems.

The numerical model can also be extended to handle two-dimensional problems. Implementation of a D4 ordering scheme reduces the computer solution time and storage requirements of the model. Simulation of the migration of TCE in a confined aquifer demonstrates the applicability of the model to a field problem.

ACKNOWLEDGEMENTS

Many individuals have been instrumental in the development of this work. Thanks must be expressed to George F. Pinder who first encouraged me to pursue this project and has contributed much of his knowledge and ideas to its completion. My gratitude must also be extended to William G. Gray for his support and constructive criticisms of the text. Portions of this work have taken shape as a result of discussions with Michael A. Celia, whose contributions are also gratefully acknowledged.

Thanks should go to Elizabeth Kaminski for her excellent and creative layout and typing of this text and to Thomas Agans for his fine draftsmanship on a number of the figures.

This work was supported, in part, by the U.S. Department of Energy under contract #DE-AC02-79EV10257 and the Industrial Support Group of Princeton University. Other funding was supplied by fellowships from the DuPont Corporation, the Shell Oil Foundation, and WAPORA, Inc.

Table of Contents

CHAPTER I

INTRODUCTION

Groundwater has long been one of the world's most important resources. It accounts for approximately 96% of all fresh water in the United States and supplies more than 50% of the population with potable water. Historically, this water source has generally been regarded as pristine. However, in recent years, contamination of ground water by industrial products has become a problem of growing concern.

During the past four decades, the variety and quantity of organic chemicals produced in the U.S. has steadily increased. Currently, more than 40,000 different organic compounds are being manufactured, transported, used and eventually disposed of in the environment (Wilson, et al (1981)). Production and consumption of petroleum products has also risen in this same time period. Many of these industrial compounds are highly toxic and slightly water soluble. Thus, they pose a potential threat to large volumes of groundwater if they are somehow introduced into the subsurface. Increased production of chemicals implies the increased risk of accidental spills or leakage to the soil, and indeed, the literature abounds with contamination case histories.

Incidences of petroleum contamination of groundwater have been documented by many authors. For example, see: Schwille (1967); Toms (1971); Guenther (1972); McKee, et al (1972); Williams and Wilder (1971); Vanloocke, et al (1975); Osgood (1974); Holzer (1976): Geraghty and Miller, Inc. (1979); Yazicigil and Sendlein (1981). These case histories attest to the widespread nature and long persistence of the hydrocarbon contamination problem. In more recent years, cases of organic chemical pollution of groundwater by non-petroleum sources have also started appearing in the literature. As improved field and laboratory techniques for the detection and identification of organic pollutants are developed and as more water supplies are examined, the number of these cases will, doubtless, continue to grow. Already, organic chemicals have been found in groundwater supplies in many locations throughout the U.S. at levels exceeding EPA suggested no-adverse-response levels (SNARL's) (Weimar (1980)). For a more detailed discussion of organic pollution histories see: Lindorff (1979); Roux and Althoff (1980); Woodhull (1981); Guerrera (1981); Petura (1981); Zoeteman, et al (1981).

Many of the above articles emphasize the difficulties and costs involved in treating a contaminated groundwater supply. The removal of a chemical spill from the subsurface or the clean-up of a polluted aquifer is a complex, if not virtually impossible, task. Due to the low velocities of groundwater (on the order of 0.001 - 1 cm/sec), migration of a contaminant will generally be very slow, but this does not imply that a pollution problem is easily solved or controlled. Because chemical solubilities are low and degradation rates tend to be small, a spill may serve as an essentially unlimited contamination source to the groundwater system. If the affected soil zone is of any reasonable size, excavation of the contaminated soil is infeasible and flushing or pumping of the chemical from the soil is impossible due to the capillary forces which hold the contaminant in place. Because there are no "quick fixes" in a chemical contamination situation, it becomes important to study the processes involved in an effort to predict the movement of a pollutant and its potential impact on an aquifer system.

The infiltration and migration of a chemical contaminant in the subsurface is a complex process. Consider the hypothetical petroleum contamination scenario depicted in Figure 1.1. This cross-sectional schematic is illustrative of the important stages in the evolution of the process.

Initially, a pollutant consisting of one or more chemical species will enter the soil as the result of a chemical spill or tank leakage. It may be introduced into the subsurface as the soluble portion of an aqueous phase or as is more customarily encountered, a distinct non-aqueous phase. Under the influence of gravity, the pollutant will migrate downwards through the unsaturated zone as a separate phase. Due to capillary forces, this vertical infiltration will also be accompanied to some extent by a lateral migration. A gaseous envelope of chemical vapor may also extend beyond the main body of contamination due to the volatilization of light components in the polluting phase.

If the spill is sufficient in size, some of the contaminant will eventually reach the vicinity of the water table. Here, soluble components will dissolve, forming a "plume" of contaminated water extending outwards from the main zone of contamination. This plume may then migrate as part of the groundwater system. That portion of the contaminant

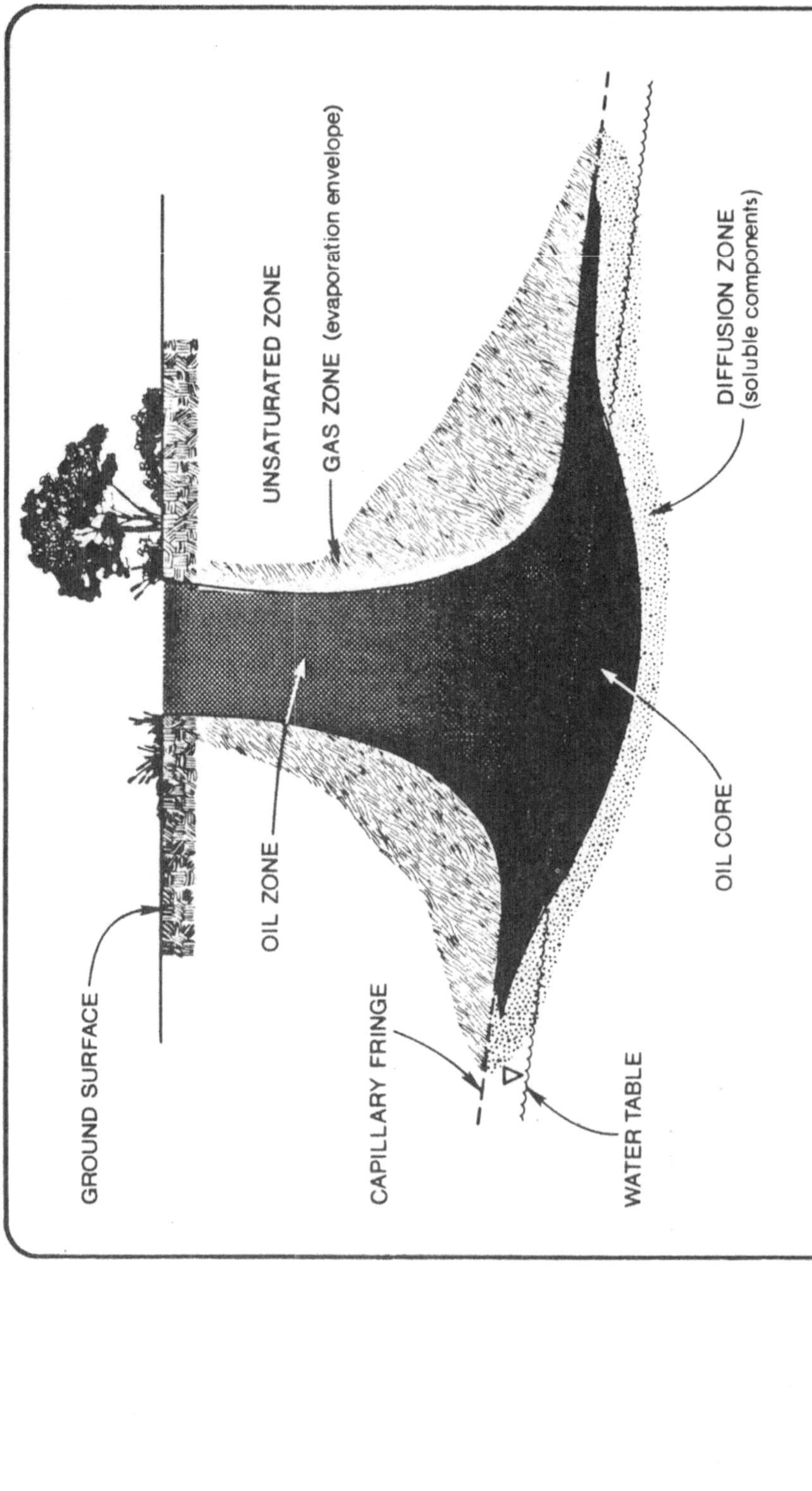

AN OIL SEEPAGE SCHEMATIC
FIGURE 1·1

which remains as a distinct phase will travel under its own pressure gradients in the capillary fringe zone and may depress natural groundwater levels.

Note that much of the non-aqueous phase will remain trapped in the unsaturated zone for an indefinite time period due to the action of capillary forces. This trapped phase may serve as a source of contamination to infiltrating rain water or a rising water table. The gaseous envelope may also function in the same manner, dissolving its chemical components in the infiltrating water.

Some natural processes may serve to retard the migration of the pollutant. These processes include biodegradation of individual chemical components by soil microorganisms and the adsorption of various components onto soil particles. Biodegradation in the groundwater zone, however, generally proceeds very slowly due to the zone's anaerobic character.

Mathematical models have been developed in the effort to quantify some of the phenomena described above. Many laboratory and field studies have also been performed to aid in the understanding of these processes and to help in model verification. See, for example, the careful experimental work of Mull (1969), Kleeberg (1969), Sprenger (1969), Schwille (1971), and Schiegg (1980). Most of these experiments and models were designed to assess the impact of petroleum product contaminants on the subsurface, since this problem has been recognized for a much longer time than that of other types of chemical contamination. The mathematical models may be grouped roughly into two categories: those that deal with the spread of the contaminant plume of dissolved components and those that deal with the movement of a chemical contaminant as a distinct phase.

A very simple model, designed by Strack (1977), falls into the first category. This model computes the optimum locations for drains which could intercept a hydrocarbon-contaminated groundwater stream. Assuming that a spill takes the form of a fixed oil zone or "pancake" of oil lying on the water table, the model determines a drain position through which all flow lines from the pancake must pass. In the mathematical formulation, three simple flow solutions for an infinite aquifer are superimposed to obtain a set of three equations which are solved iteratively. The two-dimensional cross-sectional model developed is for steady-state flow (constant aquifer recharge rate), a horizontal water table situation and a homogeneous, isotropic aquifer. Because the model cannot calculate pollutant concentrations nor deal with spatial or temporal variability, it has very limited applicability.

Other models have been developed based on this same concept of an immovable pancake of oil in the capillary fringe zone. Hoffmann (1969, 1971) treats this pancake as a pollution source whose strength is proportional to what he terms the "face of contact," an area dependent upon fluid saturations in the soil. The constant of proportionality he calls an "exchange constant". This parameter is dependent on the two fluids being modeled. The movement of a soluble contaminant from the source is described by a convective-dispersive equation for a uni-directional flow field. Both the water velocity and coefficient of dispersion are assumed to be spatially uniform and known. Any losses of solute due to adsorption or biodegradation are neglected. Hoffman proposes that the oil pancake may be treated as a point source of constant strength if one is interested in obtaining concentrations a distance from the source. This approximation enables him to use an analytical solution to the convective-dispersive equation to predict solute concentration as a function of time and distance from the source.

In Hoffmann's analysis, an abrupt boundary between the oil and water phases is assumed. This is not the case, however, in real groundwater situations where the presence of an initial soil moisture profile introduces multiple contact interfaces. During infiltration, the contact area will actually change continuously in both geometry and size. As explained above, Hoffmann's model is not applicable to the determination of concentration profiles in the vicinity of the source since the point source assumption will no longer be valid in this area. His approach has merit, however, in the determination of the extent of contaminant migration in a very simple flow field if the size and characteristics of the spill can be estimated.

Fried, et al (1979) and Bastien, et al (1977) also applied Hoffmann's methods to the solute transport problem. They use a different equation to calculate source strength which is based on their experimental findings. Mass pollution rate is given as a function of fluid velocity and equilibrium concentration. The steady-state solution to the convective-dispersive equation is examined to determine the maximum areal extent of the plume for a given threshold concentration.

Other more complex theories which incorporate additional equation terms for adsorption or decay have been used extensively in the soil science literature to model transport of solutes. Most applications have been in the areas of soil nutrients and pesticides. See Boast (1973) for a discussion of the various forms of the convective-dispersive equation which have been used. Analytical solutions to some of these equations are presented by van Genuchten and Wierenga (1976). In their model for soluble oil transport, Duffy, et al (1977) include a linear adsorption isotherm and a first-order decay term. Enfield, et al (1982) develop three different mathematical models for the evaluation of transport of organic pollutants. They consider non-linear adsorption in addition to degradation.

The mathematical models described in the previous paragraphs are primarily simplified analytical approaches to the complex problem of solute transport. In the effort to incorporate aquifer heterogeneities, many modelers have turned instead to numerical solution methods.

A numerical implementation of the streamline approach has been used to predict time of travel for dissolved hydrocarbons in a contaminated aquifer in Iowa (Yazicigil and Sendlein (1981)). Transient head distributions in the aquifer are calculated using a two-dimensional numerical aquifer flow model. From these distributions and hydraulic conductivity values, velocities are computed and streamlines drawn. Travel times are estimated from this information. Only convective movement of the contaminant is considered; no dispersive movement or mass loss by adsorption or biodegradation is included.

Other modelers have taken a more sophisticated approach to the numerical modeling of solute transport. Many models have been developed which couple the solutions of a groundwater flow equation and the convective-dispersive equation to obtain distributions of contaminant concentrations in time. See for example: Pinder (1973); Konikow and Bredehoeft (1978); Pickens and Grisak (1979); Gray and Hoffman (1983 a,b). These models deal essentially with far-field contaminant migration; the actual mechanisms of dissolution to form a contaminant plume are not treated.

The migration of contaminants in the unsaturated zone is also a subject which has been receiving a great deal of attention in recent years. Numerical simulation of solute transport in this zone involves the coupling of the solutions of the unsaturated mass flow equation and a convective-dispersive equation. Examples of this type of numerical approach may be found in the work of Bresler (1973), Duguid and Reeves (1976), Segol (1977), or van Genuchten (1982). Again, the mechanisms of dissolution to form the contaminant plume are not considered.

Van Dam (1967) presented the first detailed analysis of hydrocarbon pollution of groundwater as a two-phase problem. He incorporates capillary pressure in his expression of fluid potential and discusses the various stages of the infiltration process based on this potential. These stages include downward seepage and horizontal spreading of the oil phase along the water table. An equation for the lateral extent of oil migration is developed, based on the average height of the oil-gas capillary zone and the concept of a residual (immobile) saturation. This equation can provide a rough estimate of the limits of oil migration as a distinct phase.

Mull (1969, 1971, 1978) also analyzed the infiltrating oil problem. Neglecting capillary forces, he bases his equation for limited oil infiltration on the concept of "piston flow". Moisture profiles are simplified as rectangles such that saturation is inversely proportional to distance from the source. Using Darcy's law to express the velocity of the oil, he is able to develop an integrable expression for distance of migration versus time. Migration is assumed to cease when oil reaches a residual level. An expression relating the maximum depression of the water table (due to the oil body) to the oil volume and fluid densities is also developed. Dracos (1978) used Mull's assumptions to derive similar expressions for infiltration distance versus time for an area source, line source, and point source.

Schiegg (1977b) extended the basic idea of piston flow to the non-horizontal water table problem. Under this condition, oil spread will be nonsymmetric, having a greater lateral extent in the direction of groundwater flow. Schiegg divides the infiltration of oil into two phases: vertical migration and spreading on the water table. Analytical methods are used to calculate the total time and maximum radius of extent for the first phase. Assuming that, at the completion of the first phase, the oil body is in the form of a cylinder of known radius, he then uses a computer program to calculate the time to completion of migration. The maximum dimensions of the oil pancake are also computed.

In all of the models discussed above, the soil is assumed homogeneous and isotropic. Only the displacement of water and air by oil can be considered and capillary pressure between the oil and water phases is neglected. These types of models are useful only in the estimation of the area of spill influence as a function of time. Because of the many assumptions they entail, the estimates may not agree very well with real values, especially in situations where capillary forces play a dominant role. Mathematical development of these models tends to be very cumbersome under the weight of all the simplifying approximations and one can often lose sight of the basic balance laws.

Holzer (1976) examined the unsteady decay of an oil lens on the water table subject to water recharge from the surface. He adapted Hantush's theory for growth of a freshwater lens in a saline unconfined aquifer. A static water table and homogeneous, isotropic aquifer are assumed. The solution gives an expression relating depth of lens to time. This theory is only very approximate because any effects of capillarity (including residual saturations or relative permeabilities) are neglected. The rate of lens decay would be greatly reduced by these factors. Also, this analysis does not enable one to compute the lateral extent of the lens. In a real situation, the problem is further complicated by the fact that recharge water must pass through an unsaturated zone which contains at least a residual level of oil.

A more rigorous approach to the two-phase flow problem was adopted by Hochmuth (1981). He developed a two-dimensional numerical model to simulate the horizontal spread of an oil body after reaching the water table. Use of an areal model necessarily implies that any vertical gradients are assumed negligible. Based on this assumption of vertical equilibrium and the concept of fluid potential, two equations are written expressing the elevations of the air/oil interface and the oil/water interface as functions of fluid densities and capillary pressures. As with many of the other models discussed previously, the capillary pressure-saturation relation is represented as a step function (piston flow). Thus, the oil phase and water phase domains are assumed to have fixed saturations; only the extent of these domains may change with time. The non-linear Boussinesq equation for flow in an unconfined aquifer is used to describe the flow of each of the phases. These two partial differential equations are discretized by a Galerkin finite element approximation in space and a finite difference approximation in time. The resulting non-linear algebraic equations are linearized by lagging interface elevations in time, and the two systems of flow equations (one for each flow regime) are then solved simultaneously by an iterative procedure.

This model, unlike those discussed previously, can handle spatial nonhomogeneity of the aquifer. The analysis is straightforward and varying boundary conditions may be employed on the flow domain. Simultaneous flow of two phases with varying saturations, however, cannot be simulated. Also, extension to a contaminating phase more dense than water is not easily handled because in that situation, the vertical dimension would necessarily become important and a distinct interface between the two fluids would not be found.

* * *

The last few pages have highlighted many of the inadequacies of the past modeling efforts in the chemical contamination problem. No single model has been developed which is capable of tracing the multiphase migration of a pollutant (i.e. its migration as a solute, a gas and a non-aqueous fluid phase). Mechanisms of interphase mass transfer have not been adequately treated, and migration of the pollutant in the gas phase has, in fact, been ignored entirely. Use of the piston flow assumption by many modelers neglects the effect of capillary forces on the distribution of fluids in the porous medium. All of these factors point to the need for a more comprehensive approach to the modeling of the chemical contamination process. This approach should enable one to determine both fluid saturations and pollutant concentrations as functions of space and time in a heterogeneous porous medium.

The purpose of this thesis is to develop such an approach and to implement it in the construction of a numerical simulator. Chapter II contains a systematic development of the model's governing equations from basic principles. A one-dimensional numerical model based on these governing equations is described in Chapter III. Application of this model to two different types of chemical contaminants is discussed in Chapter IV. Chapter V deals with the extension of this simulator to two dimensions.

CHAPTER II

EQUATION DEVELOPMENT

In this chapter, the equations governing the multiphase flow and transport of contaminants in a porous medium are derived. A general mass balance expression is developed in Section 2.1. Sections 2.2 - 2.7 deal with the application of this balance law to each of the porous medium components in turn. Based on the physical characteristics of these components, various constitutive relations and approximations are introduced. Incorporation of these constitutive relations into the balance laws make the resulting system of mass balance equations more amenable to solution. An equation summary is given in Section 2.8 .

2.1 Presentation of the Balance Laws

Consider the section of a porous medium depicted in Figure 2.1. It contains all of the elements which may be present in a typical contamination scenario as discussed in Chapter I. Four different types of regions or phases are visible: solid (soil), water, gas, and contaminant. There are distinct interfaces separating these phases and each phase maintains its own characteristics or prceerties. Note that phase separation does not preclude the possibility of rass transfer across phase boundaries

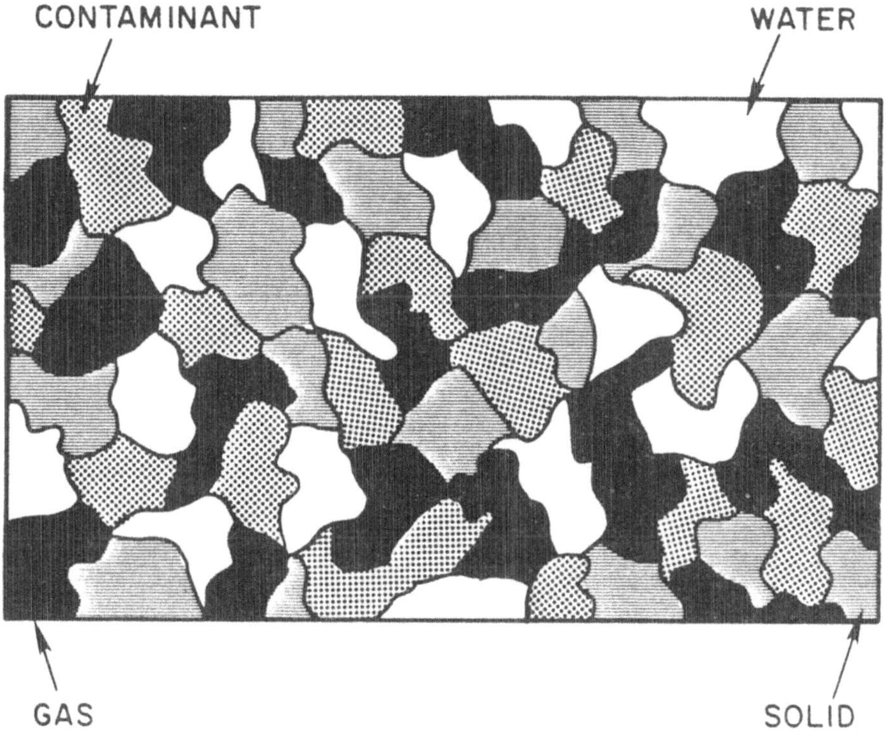

FIGURE 2.1: POROUS MEDIUM CROSS-SECTION

(such as between the contaminant and water phases) or the possibility that each phase may be formed from a number of components.

It is assumed that each subregion or phase may be represented as a continuum and that within each phase, the classical microscopic balance laws of continuum mechanics will hold. Using the microscopic mass balance law as a starting point, a macroscopic mass balance law may be formulated by application of volume averaging techniques. An outline of this procedure and the requirements for its validity are given in Appendix A. For a more complete development, see Hassanizadeh and Gray (1979 a,b).

The macroscopic mass balance law for a species i in a phase α may be written as:

$$\frac{\partial}{\partial t}(\rho_\alpha \omega_i^\alpha) + \underset{\sim}{\nabla} \cdot (\rho_\alpha \underset{\sim}{v}^\alpha \omega_i^\alpha) - \underset{\sim}{\nabla} \cdot \underset{\sim}{J}_i^\alpha - \rho_\alpha f_i^\alpha$$

$$= \rho_\alpha (e^\alpha(\rho\omega_i) + I_i^\alpha) \tag{2.1}$$

where:

ρ_α	is the mass of phase α per unit volume of the entire medium
$\underset{\sim}{v}^\alpha$	is the mass average of the α phase velocity
ω_i^α	is the mass fraction of species i in the α phase
$\underset{\sim}{J}_i^\alpha$	is an average flux vector which represents the non-advective flux of species i in the α phase
f_i^α	is the external supply of species i to the α phase
$e^\alpha(\rho\omega_i)$	represents the exchange of mass of species i due to phase change
I_i^α	represents the exchange of mass of species i due to interphase diffusion.

Note that this equation is identical to equation (A.5) in Appendix A. In (2.1), however, averaging symbols have been omitted for convenience.

Equation (2.1) is subject to the following constraints:

1. The sum of all species mass fractions over a phase α should equal unity:

$$\sum_i \omega_i^\alpha = 1 \tag{2.2}$$

   This constraint arises directly from the definition of mass fraction.

2. The total flux of all species to (or from) the α-phase is equal to the mass gained (or lost) by that phase:

$$\sum_i [e^\alpha(\rho\omega_i) + I_i^\alpha] = e^\alpha(\rho) \tag{2.3}$$

3. The total mass of the system is conserved:

$$\sum_\alpha e^\alpha(\rho) = 0 \tag{2.4}$$

4. The mass of a single (non-reacting) species is conserved over the entire system:

$$\sum_\alpha [e^\alpha(\rho\omega_i) + I_i^\alpha] = 0 \tag{2.5}$$

Summing equation (2.1) over all phases and incorporating constraint (2.5) yields a mass conservation equation for a species i :

$$\sum_{\alpha} \left\{ \frac{\partial}{\partial t} (\rho^{\alpha} \varepsilon_{\alpha} \omega_i^{\alpha}) + \underset{\sim}{\nabla} \cdot (\rho^{\alpha} \varepsilon_{\alpha} \underset{\sim}{v}^{\alpha} \omega_i^{\alpha}) - \underset{\sim}{\nabla} \cdot \underset{\sim}{J}_i^{\alpha} - \rho^{\alpha} \varepsilon_{\alpha} f_i^{\alpha} \right\} = 0 \qquad (2.6)$$

Note that in (2.6), the volume averaged bulk density of the α phase, ρ_{α}, has been replaced by the product $\varepsilon_{\alpha}\rho^{\alpha}$. Here, ε_{α} is the void fraction occupied by the α phase and ρ^{α} is an intrinsic volume average of mass density. This substitution is merely for convenience; it expresses density in its more commonly encountered form.

Equation (2.6) is subject to constraint (2.2) and also to the constraint:

$$\sum_{\alpha} \varepsilon_{\alpha} = 1 \quad , \qquad (2.7)$$

from the definition of void fraction as volume of phase/total volume. Mass conservation equations for all the species present in the system will be developed from (2.6).

It is necessary at this point in the development to determine the precise number and type of species to be considered. It is assumed that the organic contaminant phase (hereafter referred to as *phase o*) can consist of, at most, two components or species (hereafter referred to by subscripts 1 and 2). These two species may have distinct chemical formulae or they may represent groups of compounds with similar properties. A particular property of a species would, in the latter case, be an averaged property of the group of compounds. One of the organic pecies, species 1, is assumed inert, i.e. non-reactive with the other hases. Species 2 may be a more volatile component, capable of crossing hase boundaries. Other species in the system include: soil, air, and 'ater (hereafter designated by subscripts s, a, w). Each of these com- onents is assumed to exist in one phase only -- solid, gas, and water, 'espectively (hereafter designated as s, g, w). This implies that there s no interphase mass exchange of these components. Under these assump- .ions, formation and migration of water vapor is not included in the inalysis, nor is adsorption of water onto soil particles. Thus, five listinct species have been identified and five equations of the type (2.6) iay be written to describe the physical system.

?.2 Soil Species Equation

Consider the soil species. By assumption, it is present only in :he solid phase. Equation (2.6) may, thus, be written for the soil spe- :ies as:

$$\frac{\partial}{\partial t}(\rho^s \varepsilon_s \omega_s^s) + \underset{\sim}{\nabla} \cdot (\rho^s \varepsilon_s \underset{\sim}{v}^s \omega_s^s) - \underset{\sim}{\nabla} \cdot \underset{\sim}{J}_s^s - \rho^s \varepsilon_s f_s^s = 0 \qquad (2.8)$$

The only mass exchange which may occur across a solid-fluid interface is that which is due to the adsorption of organic species 2 onto the soil grains. If the mass of adsorbed organic is very small in comparison to the mass of the soil, as an approximation, this adsorbed mass may be neglected in the soil equation. This approximation may be written as: $\omega_s^s \simeq 1$.

Because there is neither internal production nor non-advective movement of the soil component, the macroscopic mass balance equation (2.8) may be simplified as:

$$\frac{\partial}{\partial t}(\rho^s \varepsilon_s) + \underset{\sim}{\nabla} \cdot (\rho^s \varepsilon_s \underset{\sim}{v}^s) = 0 \qquad (2.9)$$

This equation may be manipulated algebraically to yield:

$$\varepsilon_s \frac{D^s \rho^s}{Dt} + \rho^s \frac{\partial \varepsilon_s}{\partial t} + \rho^s \underset{\sim}{\nabla} \cdot \varepsilon_s \underset{\sim}{v}^s = 0 \qquad (2.10)$$

where $\frac{D^\alpha}{Dt}(f) = \frac{\partial}{\partial t}(f) + \underset{\sim}{v}^\alpha \cdot \underset{\sim}{\nabla}(f)$ is the substantial derivative.

Although the soil matrix itself may deform, the individual soil grains will be regarded as rigid or incompressible. Since the mass of adsorbed organic has been neglected, this implies that $\frac{D^s \rho^s}{Dt} = 0$. Equation (2.10) now becomes:

$$\frac{\partial \varepsilon_s}{\partial t} = - \underset{\sim}{\nabla} \cdot \varepsilon_s \underset{\sim}{v}^s \qquad (2.11)$$

Let ε represent the total void fraction of all fluid phases (o, w, g). Then, from constraint (2.7), $\varepsilon_s = 1 - \varepsilon$ and (2.11) may be rewritten as:

$$\frac{\partial \varepsilon}{\partial t} = \underset{\sim}{\nabla} \cdot (1 - \varepsilon) \underset{\sim}{v}^s \qquad (2.12)$$

This equation provides an expression for the change in total void fraction with time and will be incorporated into the other species equations.

If an isothermal system is assumed, deformation of the soil matrix (and consequent change in void fraction) will be caused solely by changes in intergranular stress (σ'). The actual mechanisms of this deformation may be very complex and will not be discussed here (see Sowers and Sowers (1970)). If lateral movement of the matrix is neglected, vertical strain may be equated with volume strain and an expression may be developed for matrix compressibility. The coefficient of compressibility, α , may be defined as:

$$\alpha = -\frac{1}{V_T}\frac{dV_T}{d\sigma'} \tag{2.13}$$

where V_T is the bulk volume of the porous medium. If it is further assumed that there is no change in the overburden (vertical) load on the soil, then any change in intergranular stress will be due to an equal and opposite change in the fluid pore pressure. Recalling that the volume of soil grains has been assumed fixed (incompressibility assumption), (2.13) may be rewritten as:

$$\alpha = -\frac{1}{\varepsilon_s}\frac{d\varepsilon_s}{dP_{ave}} \tag{2.14}$$

where P_{ave} is an average soil fluid pressure defined in Section 2.3 (equation (2.30)). Note that α will have a non-negative value.

In general, α will be neither constant nor a unique function of the applied stress. Its value will actually depend on the loading history and void fraction of the soil, in addition to the soil type. For small changes in void fraction, however, it will be assumed that the stress-strain curve for the medium may be approximated by a straight line. α will be regarded as a constant over this range. It will also be assumed that the soil matrix behaves elastically, i.e. that any void fraction change is reversible. For most real soils, much of the void ratio change may actually be irreversible (see Bear (1979)). This factor becomes important when considering cases of large deformations such as land subsidence. For the contamination scenario under discussion, however, deformations will tend to be very small and the elasticity assumption is reasonable.

The time derivative in (2.10) may now be expanded in terms of the coefficient of compressibility to yield:

$$\frac{\partial \varepsilon_s}{\partial t} = \frac{d\varepsilon_s}{dP_{ave}} \frac{\partial P_{ave}}{\partial t} = -\varepsilon_s \alpha \frac{\partial P_{ave}}{\partial t} \qquad (2.15)$$

This equation will be employed later in the development of the fluid equations.

2.3 Water Equation

Consider next the mass balance equation for water. In the contamination problem, the movement of water as a distinct phase is of interest. Thus, a phase equation rather than a species equation will be employed. A mass balance equation for the water phase may be obtained by summing the equations for all species present in that phase. Equation (2.1) for the water species is given as:

$$\frac{\partial}{\partial t}(\rho^w \varepsilon_w \omega_w^w) + \underset{\sim}{\nabla} \cdot (\rho^w \varepsilon_w \underset{\sim}{v}^w \omega_w^w) - \underset{\sim}{\nabla} \cdot \underset{\sim}{J}_w^w = 0 \qquad (2.16a)$$

Note that there is no mass exchange of water with the other phases nor an external supply of water to the system. A balance equation for species 2 in the water phase may be written as:

$$\frac{\partial}{\partial t}(\rho^w \varepsilon_w \omega_2^w) + \underset{\sim}{\nabla} \cdot (\rho^w \varepsilon_w \underset{\sim}{v}^w \omega_2^w) - \underset{\sim}{\nabla} \cdot \underset{\sim}{J}_2^w = \rho^w \varepsilon_w f_2^w$$

$$+ \rho^w \varepsilon_w (e^w(\rho\omega_2) + I_2^w) \qquad (2.16b)$$

The terms on the right hand side of equation (2.16b) account for the exchange of species 2 with the other phases and for the loss of species 2 from the system due to adsorption onto the soil grains or biodegradation.

Summing equations (2.16a) and (2.16b) yields the water phase mass balance equation:

$$\frac{\partial}{\partial t}(\rho^w \varepsilon_w) + \underset{\sim}{\nabla} \cdot (\rho^w \varepsilon_w \underset{\sim}{v}^w) = \rho^w \varepsilon_w f_2^w + \rho^w \varepsilon_w e^w(\rho) \qquad (2.17a)$$

Here constraints (2.2) and (2.3) have been employed. Note that since non-advective flux of the water species is equal and opposite in sign to the non-advective flux of organic species 2, these two terms cancel in the summation.

This analysis has been restricted to contaminants which are only slightly water soluble. Under this restriction, any changes in water properties (such as density or viscosity) due to the presence of soluble organics or the exchange of these organics with other phases will be very small and will be neglected. Thus, the terms on the right hand side of equation (2.17a) will be assumed negligible in comparison to the other equation terms. The equation may then be rewritten as:

$$\frac{\partial}{\partial t}(\rho^W \varepsilon_W) + \underset{\sim}{\nabla} \cdot (\rho^W \varepsilon_W \underset{\sim}{v}^W) = 0 \qquad (2.17b)$$

Expansion of (2.17b) yields:

$$\rho^W \varepsilon \frac{\partial s_W}{\partial t} + s_W \varepsilon \frac{D^W \rho^W}{Dt} - \rho^W s_W \frac{\partial \varepsilon_S}{\partial t} + \rho^W \underset{\sim}{\nabla} \cdot s_W (1-\varepsilon_S)(\underset{\sim}{v}^S + \underset{\sim}{v}^{WS}) = 0 \qquad (2.18)$$

where $\underset{\sim}{v}^{WS} = \underset{\sim}{v}^W - \underset{\sim}{v}^S$ is the relative velocity of the water phase with respect to the soil matrix,

and $s_W = \varepsilon_W/\varepsilon$ is the saturation of the water phase.

Incorporation of (2.11) into (2.18) and further manipulation produces the following form:

$$\varepsilon\rho^{w} \frac{D^{s}s_{w}}{Dt} + s_{w}\varepsilon \frac{D^{w}\rho^{w}}{Dt} + \rho^{w}s_{w} \left(- \frac{1}{\varepsilon_{s}} \frac{D^{s}\varepsilon_{s}}{Dt}\right) + \rho^{w} \underset{\sim}{\nabla}\cdot s_{w} \varepsilon \underset{\sim}{v}^{ws} = 0 \qquad (2.19)$$

Under isothermal conditions, the density of the water phase will be a function of water pressure, P^{w}, alone. Water compressibility β_{w} may be defined as:

$$\beta_{w} = \frac{1}{\rho^{w}} \frac{d\rho^{w}}{dP^{w}} \qquad (2.20)$$

For slightly compressible fluids, such as water, β_{w} may be assumed constant over the pressure range of interest (Aziz and Settari (1979)). Note that water density has been assumed independent of composition. The substantial derivative of water density may now be written in terms of pressure:

$$\frac{D^{w}\rho^{w}}{Dt} = \frac{d\rho^{w}}{dP^{w}} \frac{D^{w}P^{w}}{Dt} = \rho^{w}\beta_{w} \frac{D^{w}P^{w}}{Dt} \qquad (2.21)$$

The velocities of primary interest in this analysis are the relative velocities of the fluids to the solid phase. If the deformation of the solid matrix is small, velocities of the soil grains will be extremely small $(\underset{\sim}{v}^{s} \simeq 0)$. This approximation simplifies the substantial derivative expressions in (2.19) to the following forms:

$$\frac{D^{s}f}{Dt} = \frac{\partial f}{\partial t} + \underset{\sim}{v}^{s} \cdot \underset{\sim}{\nabla} f \simeq \frac{\partial f}{\partial t} \qquad (2.22)$$

and

$$\frac{D^w f}{Dt} = \frac{\partial f}{\partial t} + (\underset{\sim}{v}^{ws} + \underset{\sim}{v}^s) \cdot \underset{\sim}{\nabla} f \simeq \frac{\partial f}{\partial t} + \underset{\sim}{v}^{ws} \cdot \underset{\sim}{\nabla} f \qquad (2.23)$$

Incorporation of (2.20), (2.21), (2.22), and (2.23) into (2.19) yields:

$$\varepsilon \frac{\partial s_w}{\partial t} + s_w \beta_w \varepsilon \frac{\partial P^w}{\partial t} + s_w \beta_w \varepsilon \, \underset{\sim}{v}^{ws} \cdot \underset{\sim}{\nabla} P^w - \frac{s_w}{\varepsilon_s} \frac{\partial \varepsilon_s}{\partial t} + \underset{\sim}{\nabla} \cdot s_w \varepsilon \, \underset{\sim}{v}^{ws} = 0 \qquad (2.24)$$

Historically, the relative velocity of fluid motion to the soil matrix has been represented by a simple empirical law developed by Henry Darcy in 1856. Darcy experimented with the steady, incompressible vertical flow of water in homogeneous sands and found that the rate of flow was proportional to the cross-sectional area and the hydraulic gradient. The constant of proportionality is known as the hydraulic conductivity or permeability of the medium. It is a function of both the matrix and fluid properties.

Darcy's law has been extended to describe the non-steady three-dimensional flow of fluids where scalar hydraulic conductivity is replaced by a tensor. This extension of Darcy's law to deal with anisotropic, transient flows is supported by a number of theoretical analyses. In one approach, Gray and O'Neill (1976) outline the development of Darcy's equation from the averaged momentum balance equation for a fluid. They show that by neglecting convective and inertial terms (which become negligible for slow flow), the momentum equation reduces to the same form as Darcy's law.

Generalization of the Darcy equation to handle the case of multiphase flow is a conceptually attractive idea. Indeed, Darcy's law has been used extensively in its modified form in the soil science and petroleum literature to model multiphase flow in a porous medium. The validity of this approach has been supported by a number of experimental investigations. Childs and Collis-George (1950) examined unsaturated permeabilities for steady state water flows in soil columns. Work of Vachaud (1967) and Watson (1966) verified the applicability of Darcy's law with steady state permeability values to the unsaturated, transient case. Leverett and Lewis (1941) and Corey, et al (1956) conducted three phase (oil, gas, water) permeability determinations in sands and sandstones for use in oil reservoir modeling.

All of the above experimental work found hydraulic conductivities to be functions of fluid saturations for the multiphase case. Thus, the generalized Darcy equation is nonlinear. The nonlinearity has traditionally been expressed by a dimensionless relative permeability coefficient which is a function of fluid saturations. Relative permeability ranges between the values 0.0 and 1.0. The product of this relative permeability and the saturated permeability for the phase yields the hydraulic conductivity. The functional dependence of relative permeability on saturation and any hysteresis effects are discussed in more detail in Chapters III and IV.

The generalized form of Darcy's law for multiphase flow may be written as:

$$\underset{\sim}{v}^{\alpha s} = \frac{-\underset{\approx}{k}\, k_{r\alpha}}{\mu_\alpha \varepsilon\, s_\alpha} \cdot (\underset{\sim}{\nabla} P^\alpha - \rho^\alpha g\, \underset{\sim}{\nabla} z) \tag{2.25}$$

where $\underset{\approx}{k}$ is the intrinsic permeability tensor - a function of the soil matrix $[L^2]$

$k_{r\alpha}$ is the relative permeability of the matrix to the α phase $(0 \leq k_{r\alpha} \leq 1)$

μ_α is the dynamic viscosity of the α phase $[F \cdot T/L^2]$

g is the acceleration due to gravity $[L/T^2]$

z is the vertical coordinate (positive upwards).

This equation will now be introduced into (2.24) to express the relative velocity of water to soil in terms of pressure and elevation gradients. (2.24) becomes:

$$\varepsilon \frac{\partial s_w}{\partial t} + s_w \beta_w \varepsilon \frac{\partial P^w}{\partial t} - \beta_w \frac{\underset{\approx}{k}\, k_{rw}}{\mu_w} \cdot (\underset{\sim}{\nabla} P^w - \rho^w g\, \underset{\sim}{\nabla} z) \cdot \underset{\sim}{\nabla} P^w - \frac{s_w}{\varepsilon_s} \frac{\partial \varepsilon_s}{\partial t} - \underset{\sim}{\nabla} \cdot \left[\frac{\underset{\approx}{k}\, k_{rw}}{\mu_w} \cdot (\underset{\sim}{\nabla} P^w - \rho^w g\, \underset{\sim}{\nabla} z) \right] = 0 \qquad (2.26)$$

Since β_w has a very low value $(4.531 \times 10^{-11}\ \mathrm{cm}^2/\mathrm{dyne}$ at $20°\mathrm{C})$, and pressure gradients are generally small in groundwater flow (Aziz and Settari (1979)), products of these terms in (2.26) will be neglected yielding:

$$\varepsilon \frac{\partial s_w}{\partial t} + s_w \beta_w \varepsilon \frac{\partial P^w}{\partial t} - \frac{s_w}{\varepsilon_s} \frac{\partial \varepsilon_s}{\partial t} - \underset{\sim}{\nabla} \cdot \left[\frac{\underset{\approx}{k}\, k_{rw}}{\mu_w} \cdot (\underset{\sim}{\nabla} P^w - \rho^w g\, \underset{\sim}{\nabla} z) \right] = 0 \qquad (2.27)$$

Equation (2.27) contains time derivatives in both saturation and pressure. These two variables, however, may be related through the concept of capillarity. Interfacial tension between two fluids gives rise to a curvature of the interfacial boundary. The pressure differential across the interface is called *capillary pressure*. Due to this pressure differential, one fluid, the "*wetting*" fluid, may encroach upon the region of the "*non-wetting*" fluid in a porous medium. This encroachment is known as *imbibition*. For the case of water and air, the region of encroachment is called *the capillary fringe zone*. Conversely, *drainage* is the process by which a non-wetting fluid displaces a wetting fluid.

As discussed briefly in Chapter I, capillarity has been found to play an important role in the migration of a chemical contaminant. In the case of oil pollution, there is ample evidence to support the assertion that the lateral movement of an intruding oil body takes place within the capillary fringe and is controlled by capillary forces. Capillarity is also responsible for an effect common in layered soils. In such soils, hydrocarbons have been observed to spread preferentially in more permeable layers (Schwille (1967)). This phenomena can be explained by the spatial variation of capillary properties in a layered soil and the potential gradient created by this variation. An abrupt change of soil properties can produce a "*capillary barrier*" to the flow. Dietz (1971) reported an example of this barrier effect when describing the results of an unpublished experiment by Dumore. This experiment illustrated how weak soil layering could produce major irregularities in the macroscopic fluid saturation distribution.

In view of the above discussion, it is apparent that the relationship between interface curvature and fluid saturation is of fundamental importance in the determination of the distribution of immiscible fluids in a porous medium. Leverett (1941) explored this thermodynamic relationship in detail and was able to derive a curvature-saturation function analytically for simple soil structures (such as regularly packed spheres). For real soils, however, it is necessary to use experiments to determine capillary pressure versus saturation curves for two fluids.

Capillary curves vary from medium to medium depending upon the fluids under consideration and the soil permeability, porosity, and type. They also exhibit hysteresis or multi-valuedness when the direction of saturation change has been reversed. Knowledge of the sample wetting history, however, permits one to determine saturation uniquely. Schiegg (1977a) investigated the effect of displacement dynamics on the capillary pressure versus saturation curves for the fluids air and water. He found these curves to be virtually independent of the applied gradient, closely approximating the curves determined under static conditions. A more detailed discussion of capillary curves and their extension to three phases is given in Chapters III and IV.

Based on the above discussion, water saturation will be assumed to be a known function of capillary pressure:

$$s_w = f(P_{ow}, P_{wg}) \qquad (2.28)$$

where $P_{ow} = p^o - p^w$ and $P_{wg} = p^w - p^g$ are the capillary pressures between the chemical and water phases and the water and gas phases, respectively. The partial derivative of saturation with respect to time may consequently be expanded as:

$$\frac{\partial s_w}{\partial t} = \frac{\partial s_w}{\partial P_{ow}} \frac{\partial P_{ow}}{\partial t} + \frac{\partial s_w}{\partial P_{wg}} \frac{\partial P_{wg}}{\partial t} \tag{2.29}$$

Recall that the time derivative of the void fraction can also be expanded in terms of derivatives of pressure (see equation (2.15)). Let P_{ave} be defined as a weighted average of the fluid pressures P_{og} and P_{wg}:

$$\begin{aligned} P_{ave} &= \kappa P_{og} + (1-\kappa) P_{wg} \\ &= \kappa(P_{ow} + P_{wg}) + (1-\kappa) P_{wg} \\ &= \kappa P_{ow} + P_{wg} \end{aligned} \tag{2.30}$$

where κ is a weighting parameter.

Incorporation of (2.29), (2.15), and (2.30) into the water phase mass balance equation (2.27) yields:

$$\varepsilon \left[\frac{\partial s_w}{\partial P_{ow}} \frac{\partial P_{ow}}{\partial t} + \frac{\partial s_w}{\partial P_{wg}} \frac{\partial P_{wg}}{\partial t} \right] + s_w \beta_w \varepsilon \frac{\partial P^w}{\partial t}$$

$$+ s_w \alpha \frac{\partial}{\partial t} (\kappa P_{ow} + P_{wg}) - \underset{\sim}{\nabla} \cdot \left[\frac{\underset{\approx}{k} k_{rw}}{\mu_w} \cdot (\underset{\sim}{\nabla} P^w - \rho^w g \underset{\sim}{\nabla} z) \right] = 0 \tag{2.31}$$

Numerical experimentation has shown that the solution of the system of mass balance equations is insensitive to the choice of the parameter κ . This is due to the fact that matrix compressibility α , is generally of very small magnitude. For modeling purposes, κ will be chosen as $\kappa = s_o/(s_o + s_w)$. This definition of κ is a reasonable choice because P_{ave} will reduce to the pressure of a single phase (water or organic) if only that phase is present. Note that the pressure of the gas phase has not been considered in the definition of P_{ave} . As explained in Section 2.5, the pressure of the gas phase will be assumed constant.

2.4 Inert Chemical Species Equation

The development of a mass balance equation for the inert chemical species begins with a balance law of the form (2.6). Recalling that there is no external supply of species 1 to the system and that this component is present only in the organic phase, equation (2.6) reduces to:

$$\frac{\partial}{\partial t}(\rho^o \varepsilon_o \omega_1^o) + \underset{\sim}{\nabla} \cdot (\rho^o \varepsilon_o \underset{\sim}{v}^o \omega_1^o) - \underset{\sim}{\nabla} \cdot \underset{\sim}{J}_1^o = 0 \qquad (2.32)$$

Note that because no restrictions have been placed on the amount of species 2 which may be present in the organic phase, organic phase properties may depend upon phase composition. By expanding terms and incorporating definitions of saturation and relative velocity, (2.32) may be rewritten as:

$$\omega_1^o s_o \; \varepsilon \frac{D^o \rho^o}{Dt} + \rho^o \varepsilon \frac{D^s s_o \omega_1^o}{Dt} + s_o \omega_1^o \rho^o \left(- \frac{1}{\varepsilon_s} \frac{D^s \varepsilon_s}{Dt}\right)$$

$$+ \rho^o \; \underset{\sim}{\nabla} \cdot (s_o \varepsilon \omega_1^o \; \underset{\sim}{v}^{os}) - \underset{\sim}{\nabla} \cdot \underset{\sim}{J}_1^o = 0 \qquad (2.33)$$

Equation (2.12) has also been employed in this expansion.

Following the same line of reasoning as in the development of the water equation, soil velocity will be neglected and Darcy's law will be used to express the relative velocity of organic to solid. Equation (2.33) becomes:

$$\omega_1^o s_o \varepsilon \; \frac{\partial \rho^o}{\partial t} - \omega_1^o \left[\frac{\underset{\approx}{k} \; k_{ro}}{\mu_o} \cdot (\underset{\sim}{\nabla} P^o - \rho^o g \; \underset{\sim}{\nabla} z) \cdot \underset{\sim}{\nabla} \rho^o \right]$$

$$+ \rho^o \varepsilon \; \frac{\partial s_o \omega_1^o}{\partial t} - \frac{s_o \omega_1^o \; \rho^o}{\varepsilon_s} \frac{\partial \varepsilon_s}{\partial t} - \rho^o \underset{\sim}{\nabla} \cdot \left[\frac{\omega_1^o \underset{\approx}{k} \; k_{ro}}{\mu_o} \cdot (\underset{\sim}{\nabla} P^o - \rho^o g \; \underset{\sim}{\nabla} z) \right]$$

$$- \underset{\sim}{\nabla} \cdot \underset{\sim}{J}_1^o = 0 \qquad (2.34)$$

Under isothermal conditions, the density and viscosity of the organic phase will generally be functions of both phase pressure and phase composition. Constraint (2.2) implies that knowledge of ω_1^o will define the composition of the chemical phase. Thus, density and viscosity may be expressed as:

$$\rho^o = f_1(P^o, \omega_1^o)$$

$$\mu_o = f_2(P^o, \omega_1^o) \qquad (2.35)$$

Extending the concept of compressibility to deal with this dual dependency of density, two compressibilities will be defined:

$$\beta_o^P = \frac{1}{\rho^o} \left.\frac{\partial \rho^o}{\partial P^o}\right|_{\omega_1^o}$$

$$\beta_o^1 = \frac{1}{\rho^o} \left.\frac{\partial \rho^o}{\partial \omega_1^o}\right|_{P^o} \tag{2.36}$$

Note that these compressibilities need not be constant over the pressure and composition ranges of interest. The time derivative of density may be expanded with the use of (2.36) as:

$$\frac{\partial \rho^o}{\partial t} = \frac{\partial \rho^o}{\partial P^o}\frac{\partial P^o}{\partial t} + \frac{\partial \rho^o}{\partial \omega_1^o}\frac{\partial \omega_1^o}{\partial t}$$

$$= \rho^o \beta_o^P \frac{\partial P^o}{\partial t} + \rho^o \beta_o^1 \frac{\partial \omega_1^o}{\partial t} \tag{2.37}$$

A functional dependency of saturation on capillary pressure will again be assumed. Implicit in this assumption is that this relation is independent of fluid composition. For large changes in composition, this approximation is most certainly inadequate. In most real cases, however, insufficient laboratory data is available to represent capillary relations any more completely. Making this assumption, the time derivative of saturation may be written as:

$$\frac{\partial s_o}{\partial t} = \frac{\partial s_o}{\partial P_{ow}} \frac{\partial P_{ow}}{\partial t} + \frac{\partial s_o}{\partial P_{wg}} \frac{\partial P_{wg}}{\partial t} \tag{2.38}$$

Relations (2.37) and (2.38) may be incorporated into (2.34) along with the compressibility of the soil matrix, α , to produce:

$$\rho^o \omega_1^o s_o \varepsilon \left[\beta_o^P \frac{\partial P^o}{\partial t} + \beta_o^1 \frac{\partial \omega_1^o}{\partial t} \right] + \rho^o \varepsilon s_o \frac{\partial \omega_1^o}{\partial t}$$

$$+ \rho^o \varepsilon \omega_1^o \left[\frac{\partial s_o}{\partial P_{ow}} \frac{\partial P_{ow}}{\partial t} + \frac{\partial s_o}{\partial P_{wg}} \frac{\partial P_{wg}}{\partial t} \right]$$

$$- \omega_1^o \left[\frac{\underset{\approx}{k}\, k_{ro}}{\mu_o} \cdot (\underset{\sim}{\nabla} P^o - \rho^o g \underset{\sim}{\nabla} z) \right] \cdot (\rho^o \beta_o^P \underset{\sim}{\nabla} P^o + \rho^o \beta_o^1 \underset{\sim}{\nabla} \omega_1^o)$$

$$+ s\, \omega_1^o \rho^o \alpha \frac{\partial}{\partial t} (\kappa P_{ow} + P_{wg}) - \rho^o \underset{\sim}{\nabla} \cdot \left[\frac{\omega_1^o \underset{\approx}{k}\, k_{ro}}{\mu_o} \cdot (\underset{\sim}{\nabla} P^o - \rho^o g \underset{\sim}{\nabla} z) \right]$$

$$- \underset{\sim}{\nabla} \cdot \underset{\sim}{J}_1^o = 0 \tag{2.39}$$

Note that the gradient of density in (2.34) has been expanded in a manner analogous to (2.37). For slightly compressible fluids, products of compressibilities and pressure gradients will be very small and some terms in (2.39) could be neglected. Since no restrictions have been made on the characteristics of the organic phase, however, these terms have been retained for generality.

A Fickian-type form of the non-advective flux vector, $\underset{\sim}{J}_1^o$, will now be postulated:

$$\underset{\sim}{J}_1^0 = \rho^0 \varepsilon s_0 \underset{\approx}{D}^0 \cdot \underset{\sim}{\nabla} \omega_1^0 \tag{2.40}$$

where $\underset{\approx}{D}^0$ is a macroscopic second-order tensor incorporating both diffusive and dispersive effects. A more general form of the flux vector might include terms for pressure-, forced-, and thermal-diffusive effects (Bird, et al (1960)). However, for an isothermal system and non-ionic species, the last two types of diffusion vanish. In addition, the pressure diffusion term will be negligible for pressure gradients of the magnitude present in a groundwater system.

Theoretical justification for the use of equation (2.40) may be given by constitutive theory. To approximate the functional form of the flux vector, a Taylor's series expansion, with respect to the independent constitutive variables, may be constructed about an equilibrium state $(\underset{\sim}{J}_1^0 \big|_E = 0)$. By the use of such an expansion and various physical arguments, Shapiro (1981) develops an expression for the flux vector for an isotropic system. If the higher-order terms in mass fraction gradient are neglected and an isothermal system is assumed, this expression simplifies to the form (2.40), where $\underset{\approx}{D}^0$ is given by:

$$\underset{\approx}{D}^0 = \underset{\approx}{D}^m + \underset{\approx\approx}{D}^1 : \underset{\sim}{v}^{os} \underset{\sim}{v}^{os} \tag{2.41}$$

$\underset{\approx}{D}^m$ is a second-order macroscopic diffusion tensor, incorporating molecular diffusion and tortuosity effects and $\underset{\approx}{D}^1$ is a fourth-order dispersion tensor. These tensor coefficients are spatially dependent. The above representation of $\underset{\approx}{D}^0$ is similar to the expression derived by Bear and Bachmat (1967) for an interconnected capillary tube model and also that developed intuitively by Whitaker (1967).

2.5 Air Species Equation

An equation for the mass balance of the air species could be developed from equation (2.6) in a manner analogous to that presented in the preceding section. For a contamination situation, however, the movement of the air species is of little interest to the modeler. Instead, attention is focused primarily on the movement of species 2, the volatile compound, within the gas phase. Since the gas phase is in contact with the atmosphere and subject to no pressure driving force, it seems reasonable to assume that the pressure of the gas phase remains at atmospheric pressure. This assumption is often employed in the modeling of water flow in the unsaturated zone (see, for example, Freeze (1971)). Neglecting gravitational effects, which are generally very small for gases, the velocity of the gas phase relative to the soil grains may then be assumed negligible.

The assumption of a constant gas phase pressure reduces the number of unknowns in the species mass balance equation, eliminating the need for a separate air species equation. Only two independent unknown pressures will remain. These pressures will be chosen arbitrarily as the capillary pressures P_{wg} and P_{ow}. Note also that changes in the volume of the soil matrix will be caused only by changes in the organic or water phase pressures. This is the rationalization for defining P_{ave} as a weighted sum of these pressures in equation (2.30).

2.6 Species 2 Equation

The mass balance equation for species 2 is the most complicated of the species equations. Component 2 may be present in any of the three fluid phases, migrating by convection and/or diffusive/dispersive phenomena. The species may also adsorb onto soil grains or be biodegraded and consequently lost to the system. A general equation based on (2.6) may be written for this component as:

$$\sum_{\alpha=o,w,g} \left\{ \frac{\partial}{\partial t} (\rho^\alpha \varepsilon_\alpha \omega_2^\alpha) + \underset{\sim}{\nabla} \cdot (\rho^\alpha \varepsilon_\alpha \underset{\sim}{v}^\alpha \omega_2^\alpha) - \underset{\sim}{\nabla} \cdot \underset{\sim}{J}_2^\alpha - \rho^\alpha \varepsilon_\alpha f_2^\alpha \right\} = 0 \tag{2.42}$$

Two different phenomena contribute to the source/sink terms in (2.42). One involves the biologic breakdown of the contaminant. The subject of biodegradation in the subsurface is a very complex one. A brief overview of this topic for hydrocarbon contaminants is presented by the American Petroleum Institute (1972). Petroleum components have been found to exhibit a wide range of susceptibility to microbial attack. The presence of oxygen and various nutrients is required to maintain this soil bacterial action, and the surface area of hydrocarbon-water contact has been found to be extremely important to the degradation process. Thus, it is anticipated that biological decay will be of greater significance in the upper, aerated zones of a porous medium and proceed much more slowly in the anaerobic environment of the groundwater zone.

Field evidence of the biodegradation of petroleum pollutants tends to support the above conclusions. The occurrence of biodegradation in the unsaturated zone has been reported by many authors, including Schwille (1967) and Duffy, et al (1977). For spills of large magnitude, degradation in the groundwater zone, however, has been found to be slight or undetectable (Duffy, et al (1977), Kolle and Sontheimer (1969), Bartz and Kaess (1972)). Zoeteman, et al (1981) report similar experiences with various organic chemical contamination incidents.

Various laboratory studies have also been conducted on the microbial degradation of hydrocarbons in soils. Vanloocke, et al (1975) present a detailed review of investigations in this area. They stress that, due to large variation in soil properties, contaminants, and microbial strains, extrapolation of results from one particular experiment to another is virtually impossible. Information on the kinetics of the degradation process is also scarce. Experiments by Kappeler and Wuhrmann (1978) indicate that degradation rates vary among individual components of a contaminant. Different microbial strains were found to produce different intermediate products, some of which were toxic to other strains. This same lack of quantitative data and difficulty in extrapolation is found in laboratory studies of the biodegradation of organic chemicals (Lyman, et al (1982)).

From the above discussion, there appear to be no general rules or relations which may be applied to the description of the degradation of contaminants in the subsurface. In fact, the importance of biodegradation to the evaluation of the extent of pollution might also be questioned. Until more is known about the process and its limiting factors, it seems infeasible to include biodegradation in a modeling effort. This process will not be considered further in this development.

The second contribution to the external source/sink terms in (2.42) is made by the adsorption of species 2 onto the soil grains. There is a large body of literature on the adsorptive properties of soils to various hydrocarbon and organic contaminants. (See, for example, Van der Waarden, et al (1977), Nathwani and Phillips (1977), Wilson, et al (1981), and Rao and Davidson (1979)). Adsorption of a contaminant directly from the gas phase is a process which has not been examined by experimentalists and will be neglected here ($f_2^g \simeq 0$). There is experimental evidence to suggest that adsorption of the contaminant as a non-aqueous phase is also probably not an important phenomena (Schwille (1967)). Water, the wetting phase, will preferentially surround the soil grains, separating any organic from the grain surfaces. Thus, f_2^o will be assumed negligible for modeling purposes.

Many adsorption experiments have been conducted on the water soluble fraction of the chemical pollutant. Adorptive effects tend to retard the migration of a contaminant plume and may or may not be reversible. Frequently, experimental data is fit to the Freundlich equation:

$$S = KC^{1/n} \tag{2.43}$$

where S is the amount of adsorbed chemical in μg/g of soil
C is the concentration of pollutant in the water phase in ppm
K is the adsorption coefficient
n is a constant parameter.

Use of this non-linear adsorption isotherm equation assumes local equilibrium is reached between the adsorbed and unadsorbed solute. K and n may exhibit hysteresis, depending on the direction of concentration change.

Assuming a Freundlich isotherm representation of adsorption, the external sink term $\rho^w \varepsilon_w f_2^w$ in (2.42) will be replaced by the expression:

$$- \frac{\partial}{\partial t} (\rho^s \varepsilon_s S') \tag{2.44}$$

where $S' = K'(\omega_2^w)^{1/n}$ is the amount of adsorbed solute in g/g of soil. Equation (2.42) may now be expanded as:

$$\overbrace{\frac{\partial}{\partial t}(\rho^w \varepsilon_w \omega_2^w) + \underset{\sim}{\nabla} \cdot (\rho^w \varepsilon_w \underset{w\sim}{v}^w \omega_2^w) - \underset{\sim}{\nabla} \cdot \underset{\sim 2}{J}^w + \frac{\partial}{\partial t}(\rho^s \varepsilon_s S')}^{①}$$

$$+ \overbrace{\frac{\partial}{\partial t}(\rho^o \varepsilon_o \omega_2^o) + \underset{\sim}{\nabla} \cdot (\rho^o \varepsilon_o \underset{o\sim}{v}^o \omega_2^o) - \underset{\sim}{\nabla} \cdot \underset{\sim 2}{J}^o}^{②}$$

$$+ \overbrace{\frac{\partial}{\partial t}(\rho^g \varepsilon_g \omega_2^g) + \underset{\sim}{\nabla} \cdot (\rho^g \varepsilon_g \underset{g\sim}{v}^g \omega_2^g) - \underset{\sim}{\nabla} \cdot \underset{\sim 2}{J}^g}^{③} = 0 \tag{2.45}$$

The expressions ①, ②, and ③ in (2.45) will be considered in turn.

Expression ① accounts for the mass of species 2 dissolved in the water phase. Although this species has been assumed only slightly water soluble, this soluble fraction will be of primary importance in the investigation of the extent of a pollution problem. Organic chemical concentrations in the ppb range may render a water supply impotable. Thus, it is often desirable to track the migration of this soluble component. The existence of a dissolved organic zone around an infiltrating chemical body has been confirmed by a number of field investigations. (See, for example, Schwille (1967, 1971), Kaess (1972), and Bartz and Kaess (1972)). The importance of diffusive/dispersive transport to the movement of a solute is well known. Some of the modeling work in this area has been described in Chapter I. Thus, both convective and non-advective flux terms must be retained in ①.

Expression ① may be expanded in the same manner as equation (2.33) to yield:

$$\omega_2^W s_W \, \varepsilon \frac{D^W \rho^W}{Dt} + \rho^W \varepsilon \frac{D^S s_W \omega_2^W}{Dt} + s_W \, \omega_2^W \rho^W \left(- \frac{1}{\varepsilon_S} \frac{D^S \varepsilon_S}{Dt}\right)$$

$$+ \rho^W \underset{\sim}{\nabla} \cdot (s_W \varepsilon \omega_2^W \underset{\sim}{v}^{WS}) - \underset{\sim}{\nabla} \cdot \underset{\sim}{J}_2^W + \frac{\partial}{\partial t} (\rho^S \varepsilon_S S') \qquad (2.46)$$

This expression may be simplified by the incorporation of Darcy's law (2.25) and the definition of water compressibility (2.20):

$$\rho^W \Bigg\{ \varepsilon \, s_W \frac{\partial \omega_2^W}{\partial t} + \varepsilon \, \omega_2^W \left[\frac{\partial s_W}{\partial P_{OW}} \frac{\partial P_{OW}}{\partial t} + \frac{\partial s_W}{\partial P_{Wg}} \frac{\partial P_{Wg}}{\partial t} \right]$$

$$+ \omega_2^W \, s_W \beta_W \varepsilon \frac{\partial P^W}{\partial t} - s_W \omega_2^W \frac{1}{\varepsilon_S} \frac{\partial \varepsilon_S}{\partial t} \qquad (2.47)$$

$$- \underset{\sim}{\nabla} \cdot \left[\frac{\omega_2^W \, \underset{\approx}{k} \, k_{rW}}{\mu_W} \cdot (\underset{\sim}{\nabla} P^W - \rho^W g \, \underset{\sim}{\nabla} z) \right] \Bigg\} - \underset{\sim}{\nabla} \cdot \underset{\sim}{J}_2^W + \frac{\partial}{\partial t} (\rho^S \varepsilon_S S')$$

Note that the velocity of the soil grains has been neglected and that second-order terms in β_w and $\underset{\sim}{\nabla} P^w$ have been considered negligible as in (2.27). A functional dependence of saturation on capillary pressures (2.28) has also been assumed.

Expression ② in (2.45) may be expanded in a manner analogous to (2.34) to yield:

$$\rho^o \Bigg\{ \omega_2^o s_o \varepsilon [\beta_o^P \frac{\partial P^o}{\partial t} + \beta_o^1 \frac{\partial \omega_1^o}{\partial t}] + \varepsilon\, s_o \frac{\partial \omega_2^o}{\partial t} + \varepsilon\, \omega_2^o [\frac{\partial s_o}{\partial P_{ow}} \frac{\partial P_{ow}}{\partial t} + \frac{\partial s_o}{\partial P_{wg}} \frac{\partial P_{wg}}{\partial t}]$$

$$- \omega_2^o [\frac{\underset{\approx}{k}\, k_{ro}}{\mu_o} \cdot (\underset{\sim}{\nabla} P^o - \rho^o g\, \underset{\sim}{\nabla} z)] \cdot (\beta_o^P \underset{\sim}{\nabla} P^o + \beta_o^1 \underset{\sim}{\nabla} \omega_1^o)$$

$$- s_o \omega_2^o \frac{1}{\varepsilon_s} \frac{\partial \varepsilon_s}{\partial t} - \underset{\sim}{\nabla} \cdot [\frac{\omega_2^o \underset{\approx}{k}\, k_{ro}}{\mu_o} \cdot (\underset{\sim}{\nabla} P^o - \rho^o g\, \underset{\sim}{\nabla} z)] \Bigg\} - \underset{\sim}{\nabla} \cdot \underset{\sim}{J}_2^o \tag{2.48}$$

Here compressibilities defined by (2.36) have been utilized. A Fickian-type form of the non-advective flux vectors, $\underset{\sim}{J}_2^w$ and $\underset{\sim}{J}_2^o$, will be postulated for (2.47) and (2.48).

Expression ③ deals with the fraction of species 2 present in the gas phase. The importance of the gaseous migration of volatile hydrocarbon contaminants has been demonstrated by the field and laboratory investigations of Schwille (1971) and Schwille and Vorreyer (1969) and supported by the work of Fried, et al (1979). Pelikán, et al (1978) used soil air sampling and analysis to determine the extent of groundwater contamination in case studies. Experimental work with the volatilization of organics from solutions and soils (Dilling (1977), Kilzer, et al (1979)) suggests that the flux of volatiles is controlled by macroscopic diffusion. In accordance with these findings, a Fickian-type of diffusive flux vector will be postulated for $\underset{\sim}{J}_2^g$. Recall that the convective movement of the gas phase has been neglected. Expression (3) becomes:

$$\omega_2^g \, s_g \, \varepsilon \frac{\partial \rho^g}{\partial t} + \rho^g \varepsilon \, \frac{\partial s_g \omega_2^g}{\partial t} - \frac{s_g \, \omega_2^g \rho^g}{\varepsilon_s} \frac{\partial \varepsilon_s}{\partial t} - \underset{\sim}{\nabla} \cdot \underset{\sim}{J}_2^g \qquad (2.49)$$

Since the system is isothermal and the pressure in the gas phase remains constant, the density of this phase is controlled solely by its composition. In a manner analogous to that in the discussion of compressibility for the organic phase, a gas compressibility will be defined:

$$\beta_g = \frac{1}{\rho^g} \left. \frac{\partial \rho^g}{\partial \omega_2^g} \right|_{P^g=\text{const}} \qquad (2.50)$$

The functional dependence of gas saturation on capillary pressure will also be assumed. (2.49) becomes:

$$\rho^g \left\{ \omega_2^g s_g \, \varepsilon \, \beta_g \frac{\partial \omega_2^g}{\partial t} + \varepsilon \, s_g \frac{\partial \omega_2^g}{\partial t} + \varepsilon \, \omega_2^g \left[\frac{\partial s_g}{\partial P_{ow}} \frac{\partial P_{ow}}{\partial t} + \frac{\partial s_g}{\partial P_{wg}} \frac{\partial P_{wg}}{\partial t} \right] - \frac{s_g \, \omega_2^g}{\varepsilon_s} \frac{\partial \varepsilon_s}{\partial t} \right\} - \underset{\sim}{\nabla} \cdot \underset{\sim}{J}_2^g \qquad (2.51)$$

Combining (2.47), (2.48) and (2.51) yields the mass balance equation for species 2:

$$\rho^w \Big\{\varepsilon\, s_w \frac{\partial \omega_2^w}{\partial t} + \varepsilon\, \omega_2^w [\frac{\partial s_w}{\partial P_{ow}} \frac{\partial P_{ow}}{\partial t} + \frac{\partial s_w}{\partial P_{wg}} \frac{\partial P_{wg}}{\partial t}] + \omega_2^w\, s_w \beta_w\, \varepsilon \frac{\partial P^w}{\partial t}$$

$$- \underset{\sim}{\nabla}\cdot[\frac{\omega_2^w \underset{\approx}{k}\, k_{rw}}{\mu_w}\cdot(\underset{\sim}{\nabla} P^w - \rho^w g\, \underset{\sim}{\nabla} z)]\Big\}$$

$$+ \rho^o \Big\{\varepsilon\, s_o \frac{\partial \omega_2^o}{\partial t} + \varepsilon\, \omega_2^o [\frac{\partial s_o}{\partial P_{ow}} \frac{\partial P_{ow}}{\partial t} + \frac{\partial s_o}{\partial P_{wg}} \frac{\partial P_{wg}}{\partial t}]$$

$$+ \omega_2^o s_o \varepsilon [\beta_o^P \frac{\partial P^o}{\partial t} + \beta_o^1 \frac{\partial \omega_1^o}{\partial t}] - \omega_2^o [\frac{\underset{\approx}{k}\, k_{ro}}{\mu_o}\cdot(\underset{\sim}{\nabla} P^o - \rho^o g \underset{\sim}{\nabla} z)]\cdot(\beta_o^P \underset{\sim}{\nabla} P^o + \beta_o^1 \underset{\sim}{\nabla} \omega_1^o)$$

$$- \underset{\sim}{\nabla}\cdot[\frac{\omega_2^o \underset{\approx}{k}\, k_{ro}}{\mu_o}\cdot(\underset{\sim}{\nabla} P^o - \rho^o g\, \underset{\sim}{\nabla} z)]\Big\}$$

$$+ \rho^g \Big\{\varepsilon\, s_g\, (1+\omega_2^g \beta_g) \frac{\partial \omega_2^g}{\partial t} + \varepsilon \omega_2^g\, [\frac{\partial s_g}{\partial P_{ow}} \frac{\partial P_{ow}}{\partial t} + \frac{\partial s_g}{\partial P_{wg}} \frac{\partial P_{wg}}{\partial t}]\Big\}$$

$$+ \quad \alpha(\rho^w s_w\, \omega_2^w + \rho^o s_o\, \omega_2^o + \rho^g s_g\, \omega_2^g) \frac{\partial}{\partial t} (\kappa P_{ow} + P_{wg})$$

$$- \quad \underset{\sim}{\nabla}\cdot(\underset{\sim}{J}_2^w + \underset{\sim}{J}_2^o + \underset{\sim}{J}_2^g) + \frac{\partial}{\partial t} (\rho^s \varepsilon_s\, S') \;=\; 0 \qquad (2.52)$$

Note that equations (2.15) and (2.30) have been employed.

2.7 Partitioning of Mass

An examination of the mass balance equations developed in the preceding sections (see Section 2.8 - Equation Summary), reveals that they can be represented as non-linear partial differential equations in 6 unknowns (two capillary pressures and four mass fractions). There are three mass balance equations, subject to one constraint on mass fractions. Thus, two additional constitutive relations are still required to close the system. These constitutive relations arise out of the concept of local equilibrium.

Van der Waarden, et al (1971) investigated the applicability of an equilibrium solubility approach to the modeling of hydrocarbon solute transport. Experiments were conducted on soil columns in which residual oil zones had been created. Water was introduced at the top and permitted to trickle through the length of a column. Oil concentrations in the drain water were monitored with time during this trickling process. Results of these experiments confirmed that concentrations of hydrocarbons in the drain water readily approached equilibrium solubilities, suggesting that the transfer of soluble hydrocarbons to the water phase might be modeled as a single-stage extraction process. Partition coefficients (ratio of equilibrium concentration in water/concentration in the oil phase) were determined and found to agree with standard laboratory measurements of this parameter.

Experimental investigations of Kappeler and Wuhrmann (1978) and Fried, et al (1979) support the conclusions of Van der Waarden, et al and the use of a partition coefficient to determine solute concentrations.

Their experiments found that equilibrium concentrations were reached within tens of centimeters for flow velocities in the range of typical aquifer systems. This suggests that the partition coefficient concept may be applied on the local scale, if by "local" one understands that the predicted concentrations are values which have been averaged over some small soil volume.

The assumption of local equilibrium may also be extended to describe the volatile gas fraction. Henry's law is the thermodynamic expression which relates the vapor pressure of a solute to the mole fraction of this solute in the liquid phase. The constant of proportionality between these two variables is known as Henry's "constant". For dilute solutions and constant temperature, Henry's "constant" is, indeed, constant (given a specific solute and solvent). In general, however, Henry's constant will be a function of both pressure and composition (Wark (1971)). There is a dimensionless Henry's law constant which relates the concentration of a compound in the gas phase to its concentration in the liquid phase (Lyman, et al (1982)). Thus, one may think of this dimensionless constant as a partition coefficient.

The assumption of local equilibrium along with the concept of the partition coefficient, has been used extensively and successfully by the oil industry in the compositional modeling of oil reservoirs. See, for example, the discussion of multicomponent systems in Crichlow (1977). Partition coefficients in these models are used to determine the distribution of hydrocarbon components between the oil and gas phases. Solutes in the water phase are neglected.

For this modeling effort, local equilibrium between the three fluid phases will be assumed. The following expressions will relate mass fractions of species 2 in each phase:

$$K_2^{gw} = \frac{\omega_2^g}{\omega_2^w}$$

$$K_2^{wo} = \frac{\omega_2^w}{\omega_2^o} \tag{2.53}$$

where K_2^{gw} , K_2^{wo} are partition coefficients which may be functions of composition and pressure. These coefficients must be determined empirically. Note that use of (2.53) permits the expression of the mass fractions in the water and gas phases as functions of mass fraction in the non-aqueous organic phase:

$$\omega_2^w = K_2^{wo}\,\omega_2^o$$

$$\omega_2^g = K_2^{gw}K_2^{wo}\,\omega_2^o \tag{2.54}$$

Equations (2.54) complete the mathematical description of the system.

2.8 Equation Summary

This section contains a summary of the mass balance equations developed in Sections 2.1 - 2.7. Note that some terms have been more fully expanded and all equations have been written in terms of the two capillary pressures P_{ow} and P_{wg}. The soil species equation has been incorporated into the other balance equations and will not be listed separately here.

Water Phase Equation:

$$\varepsilon \left[\frac{\partial s_w}{\partial P_{ow}} \frac{\partial P_{ow}}{\partial t} + \frac{\partial s_w}{\partial P_{wg}} \frac{\partial P_{wg}}{\partial t}\right] + s_w \beta_w \varepsilon \frac{\partial P_{wg}}{\partial t}$$

$$+ s_w \alpha \frac{\partial}{\partial t} (\kappa P_{ow} + P_{wg}) - \underset{\sim}{\nabla} \cdot (\underset{\approx}{\tau}_w \cdot \underset{\sim}{\nabla} P_{wg}) \tag{2.55}$$

$$+ \underset{\approx}{\tau}_w \cdot \beta_w \rho^w g \, \underset{\sim}{\nabla} z \cdot \underset{\sim}{\nabla} P_{wg} + \rho^w g \, \underset{\sim}{\nabla} z \cdot \underset{\sim}{\nabla} \cdot \underset{\approx}{\tau}_w = 0$$

Organic Species 1 Equation:

$$\omega_1^o s_o \varepsilon \left[\beta_o^P \frac{\partial}{\partial t} (P_{ow} + P_{wg}) + \beta_o^1 \frac{\partial \omega_1^o}{\partial t}\right] + \varepsilon s_o \frac{\partial \omega_1^o}{\partial t}$$

$$+ \varepsilon \omega_1^o \left[\frac{\partial s_o}{\partial P_{ow}} \frac{\partial P_{ow}}{\partial t} + \frac{\partial s_o}{\partial P_{wg}} \frac{\partial P_{wg}}{\partial t}\right] - \omega_1^o \underset{\approx}{\tau}_o \cdot [\underset{\sim}{\nabla}(P_{ow} + P_{wg}) - \rho^o g \, \underset{\sim}{\nabla} z]$$

$$\cdot [\beta_o^P \underset{\sim}{\nabla}(P_{ow} + P_{wg}) + \beta_o^1 \underset{\sim}{\nabla} \omega_1^o] + s_o \omega_1^o \alpha \frac{\partial}{\partial t} (\kappa P_{ow} + P_{wg})$$

$$- \underset{\sim}{\nabla} \cdot [\omega_1^o \underset{\approx}{\tau}_o \cdot \underset{\sim}{\nabla}(P_{ow} + P_{wg})] + \omega_1^o \underset{\approx}{\tau}_o \cdot \rho^o g \, \underset{\sim}{\nabla} z \cdot [\beta_o^P \underset{\sim}{\nabla}(P_{ow} + P_{wg}) + \beta_o^1 \underset{\sim}{\nabla} \omega_1^o]$$

$$+ \rho^o g \, \underset{\sim}{\nabla} z \cdot \underset{\sim}{\nabla} \cdot \omega_1^o \underset{\approx}{\tau}_o - \frac{1}{\rho^o} \underset{\sim}{\nabla} \cdot [\rho^o \varepsilon s_o \underset{\approx}{D}^o \cdot \underset{\sim}{\nabla} \omega_1^o] = 0 \tag{2.56}$$

Organic Species 2 Equation:

$$\rho^{w}\Big\{\varepsilon s_w \frac{\partial \omega_2^w}{\partial t} + \varepsilon\,\omega_2^w\Big[\frac{\partial s_w}{\partial P_{ow}}\frac{\partial P_{ow}}{\partial t} + \frac{\partial s_w}{\partial P_{wg}}\frac{\partial P_{wg}}{\partial t}\Big] + \omega_2^w s_w \beta_w \varepsilon \frac{\partial P_{wg}}{\partial t}$$

$$- \underset{\sim}{\nabla}\cdot[\omega_2^w \underset{\approx}{\tau}_w \cdot \underset{\sim}{\nabla} P_{wg}] + \rho^w g\, \underset{\sim}{\nabla} z \cdot \underset{\sim}{\nabla}\cdot \omega_2^w \underset{\approx}{\tau}_w + \omega_2^w \underset{\approx}{\tau}_w \cdot \beta_w \rho^w g\, \underset{\sim}{\nabla} z \cdot \underset{\sim}{\nabla} P_{wg}\Big\}$$

$$+ \rho^{o}\Big\{\varepsilon s_o \frac{\partial \omega_2^o}{\partial t} + \varepsilon\,\omega_2^o\Big[\frac{\partial s_o}{\partial P_{ow}}\frac{\partial P_{ow}}{\partial t} + \frac{\partial s_o}{\partial P_{wg}}\frac{\partial P_{wg}}{\partial t}\Big]$$

$$+ \omega_2^o s_o \varepsilon\,\Big[\beta_o^P \frac{\partial}{\partial t}(P_{ow} + P_{wg}) + \beta_o^1 \frac{\partial \omega_1^o}{\partial t}\Big]$$

$$- \omega_2^o \underset{\approx}{\tau}_o \cdot [\underset{\sim}{\nabla}(P_{ow} + P_{wg}) - \rho^o g\, \underset{\sim}{\nabla} z]\cdot[\beta_o^P\, \underset{\sim}{\nabla}(P_{ow} + P_{wg}) + \beta_o^1 \underset{\sim}{\nabla}\omega_1^o]$$

$$- \underset{\sim}{\nabla}\cdot[\omega_2^o \underset{\approx}{\tau}_o \cdot \underset{\sim}{\nabla}(P_{ow} + P_{wg})] + \rho^o g\, \underset{\sim}{\nabla} z \cdot \underset{\sim}{\nabla}\cdot \omega_2^o \underset{\approx}{\tau}_o$$

$$+ \omega_2^o \underset{\approx}{\tau}_o \cdot \rho^o g\, \underset{\sim}{\nabla} z \cdot[\beta_o^P\, \underset{\sim}{\nabla}(P_{ow} + P_{wg}) + \beta_o^1 \underset{\sim}{\nabla}\omega_1^o]\Big\} \qquad (2.57)$$

$$+ \rho^{g}\Big\{\varepsilon s_g (1 + \omega_2^g \beta_g)\frac{\partial \omega_2^g}{\partial t} + \varepsilon\omega_2^g\Big[\frac{\partial s_g}{\partial P_{ow}}\frac{\partial P_{ow}}{\partial t} + \frac{\partial s_g}{\partial P_{wg}}\frac{\partial P_{wg}}{\partial t}\Big]\Big\}$$

$$+ \alpha(\rho^w s_w \omega_2^w + \rho^o s_o \omega_2^o + \rho^g s_g \omega_2^g)\frac{\partial}{\partial t}(\kappa P_{ow} + P_{wg})$$

$$- \underset{\sim}{\nabla}\cdot(\rho^o \varepsilon s_o \underset{\approx}{D}^o \cdot \underset{\sim}{\nabla}\omega_2^o + \rho^w \varepsilon s_w \underset{\approx}{D}^w \cdot \underset{\sim}{\nabla}\omega_2^w + \rho^g \varepsilon s_g \underset{\approx}{D}^g \cdot \underset{\sim}{\nabla}\omega_2^g)$$

$$+ \frac{\partial}{\partial t}(\rho^s \varepsilon_s S') = 0$$

where $\underset{\approx}{\tau}_\alpha = \dfrac{\underset{\approx}{k}\, k_{r\alpha}}{\mu_\alpha}$ is known as the α-phase mobility.

The above equations are subject to the following constraint:

$$\omega_1^o + \omega_2^o = 1 \tag{2.2}$$

and also to the equilibrium relations:

$$\begin{aligned} \omega_2^w &= K_2^{wo} \omega_2^o \\ \omega_2^g &= K_2^{gw} K_2^{wo} \omega_2^o \end{aligned} \tag{2.54}$$

Recall also that:

$$s_o + s_w + s_g = 1 \tag{2.58}$$

Constraint (2.2) and the equilibrium equations (2.54) can be incorporated directly into the partial differential equations (2.55) - (2.57). The resulting system of equations may be rewritten in the simplified form:

$$\underset{\approx}{A} \frac{\partial \underset{\sim}{u}}{\partial t} + \underset{\approx}{B}(\nabla^2 \underset{\sim}{u}) = \underset{\sim}{F} \tag{2.59}$$

where $\underset{\sim}{u}$ is the vector of unknowns $(P_{ow}, P_{wg}, \omega_1^o)$; $\underset{\approx}{A}$ and $\underset{\approx}{B}$ are 3×3 nonlinear coefficient matrices with functional dependence on t, $\underset{\sim}{x}$, $\underset{\sim}{u}$ and $\underset{\sim\sim}{\nabla u}$; and $\underset{\sim}{F}$ encompasses the lower-order terms. An equation,

i, of the above system will generally be of the parabolic-type in a given variable j as long as $A_{ij}, B_{ij} \neq 0$. For negligible capillary forces and incompressible flow, however, $A_{ij} \simeq 0$ for $j=1,2$. Under these conditions, the equations become elliptic in the pressure variables. When fluid convection dominates the system, the first-order spatial derivative terms dominate second-order terms in ω_1^0. This same dominance occurs when diffusion/dispersion is negligible ($B_{ij} \simeq 0$ for $j=3$). In this situation, the equations (2.59) display hyperbolic characteristics in the mass fraction variable. Thus, despite the parabolic appearance of equations (2.59), the system may also exhibit hyperbolic or elliptic behavior.

The three partial differential equations summarized above are similar in form to those equations used by the oil industry for compositional reservoir modeling (see Peaceman (1977)). In reservoir models, however, the water soluble portion of the non-aqueous phase is neglected along with dispersive/diffusion mass flux. Pressure in the gas phase is not assumed constant in a reservoir because the system is not in direct contact with the atmosphere and is often pressurized to facilitate oil recovery.

Chapter III

Development of the 1-D Simulator

The system of nonlinear partial differential equations developed in Chapter II is not amenable to solution by analytical means. Chapter III outlines the development of a numerical model designed to solve such a system of equations for one space dimension. After a brief discussion of the oil industry's experience in solving these types of equations, a finite difference solution scheme is described in detail in Section 3.2. Section 3.3 deals with the incorporation of boundary and initial conditions and 3.4 with the evaluation of fluid and matrix properties to define the equation parameters. The technique for the solution of the resultant set of nonlinear algebraic equations is discussed in Section 3.5. A listing of the one-dimensional computer code described in this chapter has not been included in this thesis but may be found in a separate documentation.

3.1 Background

As has been mentioned previously, the equations (2.54)-(2.58) are similar in form to those used in the oil literature to model the multiphase flow of fluids in petroleum reservoirs. Experience of the oil industry with the numerical solution of these equations is considerable. For a general overview of oil industry simulation work in this area, see the recent article by Coats (1982). It is instructive here to review some of the work which is applicable to the contamination problem.

The majority of oil models have used finite difference discretization of the governing equations. This is primarily due to the simplicity of concept and ease of application of the finite difference method. Although a Galerkin finite element approach would permit a more routine incorporation of irregular geometry, this technique has not gained wide acceptance in the modeling of multiphase flows. Generally, the physical boundaries of the flow regime in an oil reservoir are neither distinct nor accurately known (Aziz and Settari (1979)) and thus, the precise representation of these boundaries has not been of great importance to the modeler. The major reason, however, that the finite element method has not become popular with the oil industry is that difficulties are often encountered with its application to immiscible flow problems. Mercer and Faust (1977) review some of these difficulties. Use of the Galerkin finite element method entails large amounts of computation time to evaluate the coefficient matrices, and numerical oscillations and stability problems often occur at the fronts. Although some of these difficulties could perhaps be surmounted, it seems reasonable, in view of oil industry experience, to model a multiphase flow problem using a finite difference approach.

Of the various types of finite difference oil reservoir models developed, the most applicable to the groundwater contamination problem and this thesis are those which take a compositional approach to the multiphase flow of fluids. These "compositional models" are designed to solve a system of component mass balance equations much like (2.55)-(2.57) subject to some equilibrium relations or equation of state. Compositional models arose out of the need to deal with oil reservoir systems in which constant oil composition and immiscible phases were not valid assumptions. Early work in this area was done by Roebuck, et al (1969). More recent examples of compositional models may be found in Coats (1980a) and Trimble and McDonald (1981).

The solution method used by many of these compositional models is known as the "simultaneous solution" (SS) method. This method was first proposed by Douglas, et al (1959) and later extended and further analyzed by Coats, et al (1967) and Sheffield (1969). It was initially applied to two- and three-phase immiscible flow problems. The basis of the SS method is to expand all saturation derivatives in the mass balance equations in terms of pressures as was done in Chapter II in developing equations (2.55)-(2.57). The resulting system of equations is then solved simultaneously for pressures. In the absence of mass exchange, the equations are coupled solely through the saturation derivatives. Presence of mass exchange introduces additional coupling in all terms dealing with mass fractions.

Oil industry experience with these equations has shown (Aziz and Settari (1979)) that the time level at which the spatial derivatives are calculated is critical to the stability of the solution. Douglas (1960) found that the "backward difference" representation of the time derivative (all derivatives evaluated at the n+1 level) results in a stable solution for problems of this type. The backward difference approximation is given by:

$$\left[\frac{\partial}{\partial t}(f)\right]^{n+1} \approx (f^{n+1} - f^{n})\frac{1}{\Delta t} \tag{3.1}$$

where the superscripts indicate the time level at which the function is evaluated. This approximation is the one most often used in oil reservoir simulators and will be employed in this thesis work.

The properties of the SS method have been examined extensively in the oil literature for the basic two- and three-phase *immiscible* flow problems. (See Appendix B.1 for definitions of these properties.) Aziz and Settari (1979) summarize some of these investigations. They explore the stability of the SS method with respect to capillary pressures and to mobilities. By linearizing the equations (treating coefficients explicitly) and neglecting gravity effects, a Fourier stability analysis may be performed on the difference equations. This analysis demonstrates that the SS method is unconditionally stable with respect to the primary variables (capillary pressures). Because the SS method treats pressures implicitly, this result is not unexpected.

Examination of these linearized equations for stability with respect to mobilities indicates that, in this case also, stability increases with the "implicitness" of the scheme; the most stable scheme is that in which the mobilities are evaluated at the n+1 time level. This same conclusion also applies to stability with respect to the saturation derivatives. Assuming that the difference equations are shown to be consistent, then stability implies convergence of the scheme according to Lax's Equivalence Theorem (see Appendix B.1).

Note that the conclusions stated above apply only to the linearized differential operators and properly posed problems (Richtmyer and Morton (1967)). Extension to the nonlinear case is not straightforward and convergence cannot be assured. Implicit SS schemes, however, have been used successfully to solve many types of nonlinear multiphase problems. Blair and Weinaug (1969) were the first to publish a fully implicit formulation. The compositional models mentioned previously also use this approach. Thus, based on the oil industry experience, an implicit simultaneous solution scheme will be employed to solve the system of equations (2.55)-(2.57). This scheme is attractive because it offers the greatest stability and has been used successfully on similar problems.

3.2 Formation of the Difference Equations

For one space dimension, equations (2.55)-(2.57) may be rewritten as:

$$\varepsilon \left[\frac{\partial s_w}{\partial P_{ow}} \frac{\partial P_{ow}}{\partial t} + \frac{\partial s_w}{\partial P_{wg}} \frac{\partial P_{wg}}{\partial t}\right] + s_w \beta_w \varepsilon \frac{\partial P_{wg}}{\partial t} + s_w \alpha \frac{\partial}{\partial t} (\kappa P_{ow} + P_{wg}) - \frac{\partial}{\partial x} \left(\tau_w \frac{\partial}{\partial x} P_{wg}\right) + \tau_w \beta_w \gamma_w \frac{\partial P_{wg}}{\partial x} + \gamma_w \frac{\partial \tau_w}{\partial x} = 0 \tag{3.2}$$

$$\omega_1^o s_o \varepsilon[\beta_o^P \frac{\partial}{\partial t}(P_{ow} + P_{wg}) + \beta_o^1 \frac{\partial \omega_1^o}{\partial t}] + \varepsilon s_o \frac{\partial \omega_1^o}{\partial t}$$

$$+ \varepsilon \omega_1^o [\frac{\partial s_o}{\partial P_{ow}} \frac{\partial P_{ow}}{\partial t} + \frac{\partial s_o}{\partial P_{wg}} \frac{\partial P_{wg}}{\partial t}] - \omega_1^o \tau_o [\frac{\partial}{\partial x}(P_{ow} + P_{wg}) - \gamma_o]$$

$$\cdot [\beta_o^P \frac{\partial}{\partial x}(P_{ow} + P_{wg}) + \beta_o^1 \frac{\partial \omega_1^o}{\partial x}] + s_o \omega_1^o \alpha \frac{\partial}{\partial t}(\kappa P_{ow} + P_{wg})$$

$$- \frac{\partial}{\partial x}[\omega_1^o \tau_o \frac{\partial}{\partial x}(P_{ow} + P_{wg})] + \omega_1^o \tau_o \gamma_o [\beta_o^P \frac{\partial}{\partial x}(P_{ow} + P_{wg})$$

$$+ \beta_o^1 \frac{\partial \omega_1^o}{\partial x}] + \gamma_o \frac{\partial}{\partial x}(\omega_1^o \tau_o) - \frac{1}{\rho^o} \frac{\partial}{\partial x}(\rho^o \varepsilon s_o D^o \frac{\partial \omega_1^o}{\partial x}) = 0 \qquad (3.3)$$

$$\rho^w \Big\{ \varepsilon s_w \frac{\partial \omega_2^w}{\partial t} + \varepsilon \omega_2^w [\frac{\partial s_w}{\partial P_{ow}} \frac{\partial P_{ow}}{\partial t} + \frac{\partial s_w}{\partial P_{wg}} \frac{\partial P_{wg}}{\partial t}] + \omega_2^w s_w \beta_w \varepsilon \frac{\partial P_{wg}}{\partial t}$$

$$- \frac{\partial}{\partial x}(\omega_2^w \tau_w \frac{\partial P_{wg}}{\partial x}) + \gamma_w \frac{\partial}{\partial x}(\omega_2^w \tau_w) + \omega_2^w \tau_w \beta_w \gamma_w \frac{\partial P_{wg}}{\partial x} \Big\}$$

$$+ \rho^o \Big\{ \varepsilon s_o \frac{\partial \omega_2^o}{\partial t} + \varepsilon \omega_2^o [\frac{\partial s_o}{\partial P_{ow}} \frac{\partial P_{ow}}{\partial t} + \frac{\partial s_o}{\partial P_{wg}} \frac{\partial P_{wg}}{\partial t}]$$

$$+ \omega_2^o s_o \varepsilon [\beta_o^P \frac{\partial}{\partial t}(P_{ow} + P_{wg}) + \beta_o^1 \frac{\partial \omega_1^o}{\partial t}]$$

$$- \omega_2^o \tau_o [\frac{\partial}{\partial x}(P_{ow} + P_{wg}) - \gamma_o][\beta_o^P \frac{\partial}{\partial x}(P_{ow} + P_{wg}) + \beta_o^1 \frac{\partial \omega_1^o}{\partial x}]$$

$$- \frac{\partial}{\partial x}[\omega_2^o \tau_o \frac{\partial}{\partial x}(P_{ow} + P_{wg})] + \gamma_o \frac{\partial}{\partial x}(\omega_2^o \tau_o)$$

$$+ \omega_2^o \tau_o \gamma_o [\beta_o^P \frac{\partial}{\partial x}(P_{ow} + P_{wg}) + \beta_o^1 \frac{\partial \omega_1^o}{\partial x}] \Big\}$$

$$+ \rho^g \Big\{ \varepsilon s_g (1 + \omega_2^g \beta_g) \frac{\partial \omega_2^g}{\partial t} + \varepsilon \omega_2^g [\frac{\partial s_g}{\partial P_{ow}} \frac{\partial P_{ow}}{\partial t} + \frac{\partial s_g}{\partial P_{wg}} \frac{\partial P_{wg}}{\partial t}] \Big\}$$

$$+ \alpha(\rho^w s_w \omega_2^w + \rho^o s_o \omega_2^o + \rho^g s_g \omega_2^g) \frac{\partial}{\partial t} (\kappa P_{ow} + P_{wg})$$

$$- \frac{\partial}{\partial x} (\rho^o \varepsilon s_o D^o \frac{\partial \omega_2^o}{\partial x} + \rho^w \varepsilon s_w D^w \frac{\partial \omega_2^w}{\partial x} + \rho^g \varepsilon s_g D^g \frac{\partial \omega_2^g}{\partial x}) = 0 \qquad (3.4)$$

Here x is some arbitrary direction and $\gamma_\alpha = \rho^\alpha g \cos\lambda$ where λ is the angle the x direction makes with the gravity acceleration vector. Note that the adsorption term present in (2.57) has been omitted in (3.4). Adsorption was included in the mass balance equations of Chapter II because, in some cases, particularly when the soil matrix has a high organic content, this effect may retard the migration of the contaminant plume in the water phase. For the trial simulations discussed in Chapters IV and V, however, insufficient experimental data was available to quantify the adsorptive properties of the soil. Thus, inclusion of adsorptive effects in the model is not warranted at this time. The remainder of this thesis deals with a simplified system in which adsorption does not play a role.

* * * *

Consider the one-dimensional physical system depicted in Figure 3.1. This system has been discretized into a number of elements. The differential equations (3.2)-(3.4) apply anywhere within this domain. In the finite difference solution method, approximate solutions to these equations are sought at specific points or nodes. Differential operators are replaced by difference operators, thereby reducing the system of differential equations to a system of algebraic equations in discreet unknowns. Three mass balance difference equations may be written at each node i of the domain.

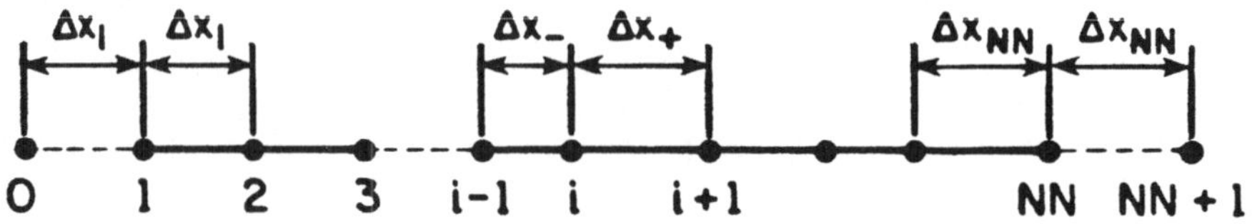

FIGURE 3.1: DISCRETIZED SYSTEM

The difference analogue to the space derivative of a function will be given as:

$$\frac{\partial}{\partial x}(f)\bigg|_i \simeq \frac{1}{\Delta x_+ + \Delta x_-}\left[f_{i+1} - f_{i-1}\right] \tag{3.5}$$

where the spatial increments Δx_+ and Δx_- are indicated in Figure 3.1. This approximation may be shown to be first-order accurate (see Appendix B.2) for unequal nodal spacing and second-order accurate for equal spacing. As $\Delta x \to 0$, the remainder term approaches zero; thus, the difference operator is consistent with the differential operator.

Consider next a differential operator of the form $\frac{\partial}{\partial x}(\Omega \frac{\partial f}{\partial x})$. This operator will be replaced by the following difference equation:

$$\frac{\partial}{\partial x}\left(\Omega \frac{\partial f}{\partial x}\right)\bigg|_i \simeq \frac{2}{\Delta x_+ + \Delta x_-}\left[\Omega_{i+\frac{1}{2}} \frac{f_{i+1} - f_i}{\Delta x_+} - \Omega_{i-\frac{1}{2}} \frac{f_i - f_{i-1}}{\Delta x_-}\right] \tag{3.6}$$

where $\Omega_{i\pm\frac{1}{2}} = \frac{1}{2}(\Omega_i + \Omega_{i\pm 1})$.

The difference operator (3.6) is also shown to be first-order accurate in Appendix B.2 and is consistent with the differential operator. It becomes second-order accurate for equal nodal spacing.

Experience of the oil industry with multiphase flow simulations has indicated that the treatment of pressure derivative terms of the form (3.6) is critical to the solution of convective dominated flow problems. These flows occur when capillary pressure, P_{ow}, is small. The centered spatial approximation for $\tau_{\alpha\, i\pm\frac{1}{2}}$ given in (3.6) has been found to produce poor solutions in problems with low diffusion (see Aziz and Settari (1979)). In these types of problems, the position of the saturation front may actually converge to a physically incorrect location. An alternative approach, known as upstream weighting, uses upstream values to compute $\tau_{\alpha\, i\pm\frac{1}{2}}$:

$$\tau_{\alpha\, i+\frac{1}{2}} = \begin{cases} \tau_{\alpha i} & \text{for flow from } i \text{ to } i+1 \\ \tau_{\alpha\, i+1} & \text{for flow from } i+1 \text{ to } i \end{cases} \tag{3.7a}$$

$$\tau_{\alpha\, i-\frac{1}{2}} = \begin{cases} \tau_{\alpha\, i-1} & \text{for flow from } i-1 \text{ to } i \\ \tau_{\alpha\, i} & \text{for flow from } i \text{ to } i-1 \end{cases} \tag{3.7b}$$

This approach has been used extensively, with good results, in oil reservoir simulations. In convection dominated problems, the upstream weighted solution has been found to converge to the true physical solution. For a good presentation of the upstream weighting concept and a discussion of its applicability, see Raithby (1976).

An upstream weighting option has been incorporated into the computer model. Within the program, $\tau_{\alpha\, i\pm\frac{1}{2}}$ is evaluated as follows:

$$\begin{aligned} \tau_{\alpha\ i+\frac{1}{2}} &= \theta_\alpha\ \tau_{\alpha i} + (1-\theta_\alpha)\tau_{\alpha\ i+1} \\ \tau_{\alpha\ i-\frac{1}{2}} &= \theta_\alpha\ \tau_{\alpha i-1} + (1-\theta_\alpha)\tau_{\alpha i} \end{aligned} \tag{3.8}$$

where $0 \leq \theta_\alpha \leq 1$ is a parameter input to the program.

Care must also be taken in the expansion of the time derivative terms to ensure that mass in the system will be conserved. Recall that the time derivative in the general mass balance equation (2.6) is given as:

$$\frac{\partial}{\partial t}\,(\rho^\alpha s_\alpha\, \varepsilon\, \omega_i^\alpha)\ .$$

This time derivative will be approximated in the model by the backward difference operator (3.1), hereafter notated as Δ_t for simplicity. Expansion of this difference operator yields the following expression:

$$\begin{aligned} \Delta_t(\rho^\alpha s_\alpha\, \varepsilon\, \omega_i^\alpha) &= (s_\alpha\, \varepsilon\, \omega_i^\alpha)^n\ \Delta_t\rho^\alpha + (s_\alpha\, \varepsilon\,)^n\ \rho^{\alpha n+1}\ \Delta_t\omega_i^\alpha \\ &\quad + s_\alpha^n(\rho^\alpha \omega_i^\alpha)^{n+1}\ \Delta_t\,\varepsilon + (\rho^\alpha\, \varepsilon\, \omega_i^\alpha)^{n+1}\ \Delta_t s_\alpha \end{aligned} \tag{3.9}$$

Algebraic manipulation will verify the time levels of the coefficients in (3.9). Expansion according to (3.9) produces what is known as a "conservative scheme."

Now consider the last term in (3.9). The backward difference operator applied to saturation may be further expanded in terms of the capillary pressures (as was done in Chapter II, Sections 2.3, 4, and 6):

$$\Delta_t s_\alpha = \frac{\partial s_\alpha}{\partial P_{wg}} \Delta t\, P_{wg} + \frac{\partial s_\alpha}{\partial P_{ow}} \Delta t\, P_{ow} = \frac{1}{\Delta t} (s_\alpha^{n+1} - s_\alpha^n) \quad (3.10)$$

For this expansion to be exact (i.e., for the above equality to hold), care must be used in the evaluation of the saturation derivatives. A chord slope derivative approximation may be expressed as:

$$\frac{\partial s_\alpha}{\partial P_{wg}} = \frac{s_\alpha^* - s_\alpha^n}{P_{wg}^{n+1} - P_{wg}^n} \qquad \frac{\partial s_\alpha}{\partial P_{ow}} = \frac{s_\alpha^{**} - s_\alpha^n}{P_{ow}^{n+1} - P_{ow}^n} \quad (3.11)$$

where s_α^* and s_α^{**} are still to be defined. Substitution of (3.11) into (3.10) yields the following equality:

$$s_\alpha^* - s_\alpha^n + s_\alpha^{**} - s_\alpha^n = s_\alpha^{n+1} - s_\alpha^n$$

or

$$s_\alpha^{**} + s_\alpha^* = 2\, s_\alpha^n + \Delta s_\alpha \quad (3.12)$$

where $\Delta s_\alpha = s_\alpha^{n+1} - s_\alpha^n$.

Suppose s_α is linear in P_{ow} and P_{wg}, that is:

$$s_\alpha = a + b\, P_{ow} + c\, P_{wg} \quad (3.13)$$

Let

$$
\begin{aligned}
s_\alpha^{*} &= s_\alpha(P_{wg}^{n+1} , P_{ow}^{n}) \\
s_\alpha^{**} &= s_\alpha(P_{wg}^{n} , P_{ow}^{n+1})
\end{aligned}
\tag{3.14}
$$

Then it may be verified that the equality (3.12) holds, and consequently, the expansion (3.10) will be exact. For the case of a nonlinear saturation function, the method (3.14) of evaluating s_α^{*} and s_α^{**} will not, in general, satisfy the equality (3.12). However, if changes in P_{ow} and P_{wg} are reasonably small, the saturation function may be represented as a linear function. Thus, the chord slope method of derivative evaluation (3.11) will conserve mass in the nonlinear case as the time step goes to zero. This chord slope approximation (equations (3.11) and (3.14)) will be used to evaluate saturation derivatives for the contaminant model.

The chord slope method could also be employed in the expansion of the time derivative of fluid density to approximate fluid compressibilities. In this case, however, the method of evaluation is not as critical to the conservation of mass. Fluid compressibilities may be regarded as weak nonlinearities (Aziz and Settari (1979)) because they are functions of properties of a single fluid and tend to change much less rapidly than the saturation derivatives. Recall also that the fractional change of porosity with pressure has previously been assumed constant over the pressure range of interest (see Section 2.2). Thus, the expansion of the time derivative of void fraction does not require special consideration here.

Expansion of the derivative terms in equations (3.2)-(3.4) according to the difference expressions (3.5)-(3.6), (3.9) yields a system of nonlinear algebraic difference equations in the unknown capillary pressures P_{ow} and P_{wg} and mass fractions ω_1^o, ω_2^o, ω_2^w, and ω_2^g. By incorporating the equilibrium relations (2.54) and constraint (2.2), this system may be reduced to a system of equations in three unknowns: P_{ow}, P_{wg}, and ω_1^o.

For a fully implicit formulation, all spatial difference terms and their coefficients would be evaluated at the n+1 time level. Some variables, however, introduce only weak nonlinearity into the system. They are functions of a single phase pressure and will not change appreciably over a time step. These variables include water and contaminant phase densities, compressibilities, and viscosities. Because the range of matrix porosity change is very small (very small compressibility), void fraction and κ will also be included in this list of parameters. In the computer model, the above variables will be "lagged" in time; they will be evaluated in the difference equations at time level n. Note that no assumptions have been made on gas phase properties. Volatilization could conceivably produce significant gas phase density changes. Thus, gas properties will be treated implicitly in the model.

In addition to those parameters discussed in the preceding paragraph, the dispersion/diffusion coefficients D^g, D^w, D^o will also be lagged in time for modeling purposes. This is equivalent to the assumption that these coefficients will vary slowly in time.

* * * * *

Let NN be the number of nodes in the discretized domain shown in Figure 3.1. Then the system of nonlinear algebraic difference equations which expresses conservation of mass over this domain may be written in matrix form as:

$$\underset{\approx}{A} \cdot \underset{\sim}{u} = \underset{\sim}{B} \tag{3.15}$$

where $\underset{\approx}{A}$ is a coefficient matrix of size $(3 \times NN) \times (3 \times NN)$, $\underset{\sim}{B}$ is a $3 \times NN$ vector and $\underset{\sim}{u}$ is an ordered vector of $3 \times NN$ unknowns.

Matrix $\underset{\approx}{A}$ is a block tridiagonal or banded matrix of bandwidth 11. Its structure is shown in Figure 3.2. Note that each group of three equations spans three nodes. The components of a block subdivision of $\underset{\approx}{A}$, $[A_{kj}]_i$ (k=1,3 j=1,9), are given in Appendix C.1 along with the corresponding values of $\underset{\sim}{B}$, $[B_k]_i$ (k=1,3). Because both $\underset{\approx}{A}$ and $\underset{\sim}{B}$ contain variables which must be evaluated at the n+1 time level, the system of equations (3.15) must be solved by some iterative method. The iterative method used in the model is discussed in Section 3.5.

3.3 Incorporation of Boundary and Initial Conditions

The coefficients tabulated in Appendix C.1 are valid for expressing the mass balance of a species at any node i of the domain. To obtain a well-posed mathematical problem, however, boundary conditions on the unknown variables are required at the endpoints as well as a set of initial conditions throughout the domain. To incorporate the boundary conditions, some modification of $\underset{\approx}{A}$ and $\underset{\sim}{B}$ must be made for the equations at the boundary nodes.

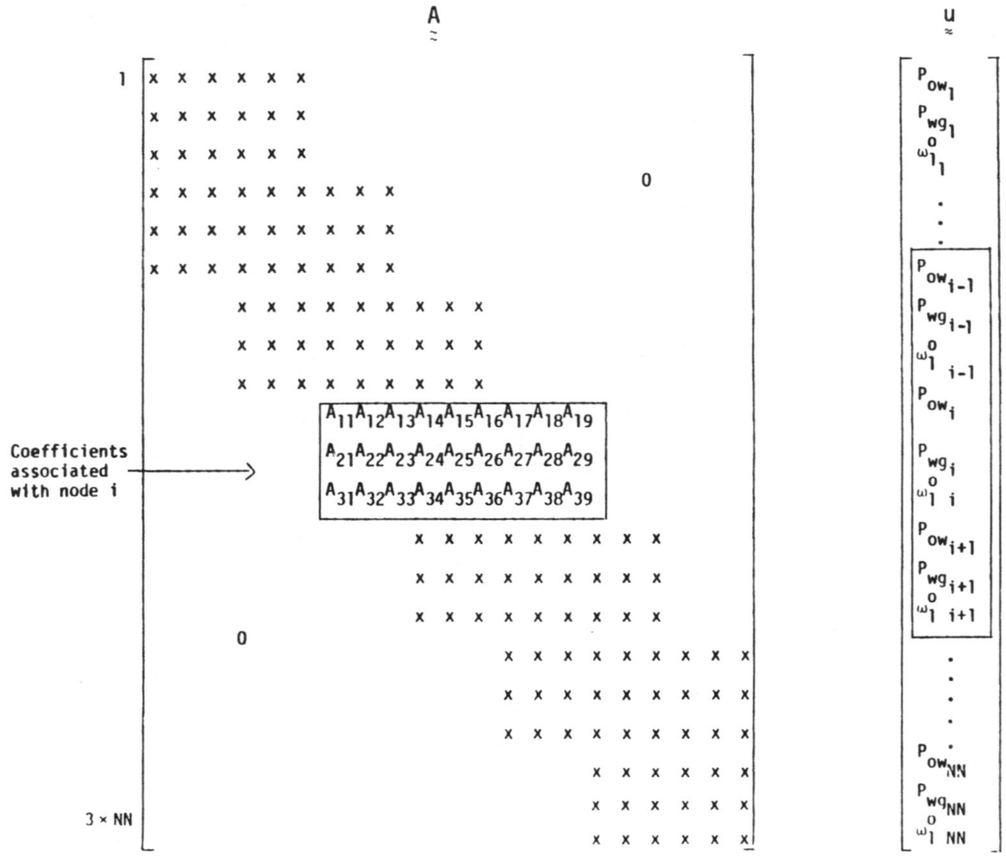

Figure 3.2: Coefficient Matrix Structure

In the case of constant pressure and composition conditions on the boundaries (commonly known as first-type or Dirichlet conditions), the modification of $\underset{\approx}{A}$ and $\underset{\sim}{B}$ is straightforward. Each mass balance equation at the boundary node i is replaced by an identity in the variables u_j of the form:

$$a_{kj} u_j \Big|_i = c_k \tag{3.16}$$

where a_{kj} and c_k are constants. In most cases, only one coefficient a_{kj} will be nonzero for each boundary equation.

For the incorporation of constant gradient conditions at a boundary node (commonly known as second-type or Neumann conditions), the full mass balance equations must be written at the node with the spatial derivatives of the unknowns replaced by the corresponding constant values. Within the computer model, this is accomplished by defining an "imaginary node" and then writing discretized equations of the same form as given in Appendix C.1 at the boundary node.

Consider the discretized domain depicted in Figure 3.1. Note that the positions of two "imaginary nodes" (nodes 0 and NN+1) have been indicated. The constant gradient of a variable u_i at node 1 may be approximated by the centered finite difference expression:

$$\frac{\partial u_i}{\partial x}\Big|_1 \simeq \frac{u_i\Big|_2 - u_i\Big|_0}{2\,\Delta x_1} = c_1 \tag{3.17a}$$

Similarly, for a second type condition at node NN, the gradient becomes:

$$\left.\frac{\partial u_i}{\partial x}\right|_{NN} \simeq \frac{u_i\big|_{NN+1} - u_i\big|_{NN-1}}{2\,\Delta x_{NN}} = c_{NN} \qquad (3.17b)$$

From (3.17) it follows that the value of u_i at an imaginary node may be expressed in terms of the boundary condition and another node in the domain:

$$u_i\big|_0 = u_i\big|_2 - 2\,\Delta x_1 c_1 \qquad (3.18a)$$

$$u_i\big|_{NN+1} = u_i\big|_{NN-1} + 2\,\Delta x_{NN} c_{NN} \qquad (3.18b)$$

These calculations are performed internally by the program. Discretized mass balance equations are then written at a constant gradient boundary node by incorporation of these imaginary node values.

In addition to boundary conditions, a set of initial conditions must be prescribed for every simulation. At each node, initial values for the three dependent variables P_{wg}, P_{ow}, and ω_1^o must be established. In areas of the domain where all three fluid phases are present, the definition of these variables is usually straightforward. Often hydrostatic pressure conditions may be employed. Consider a region, however, in which one fluid phase is absent. In such a region, the full set of mass balance equations (3.2)-(3.4) cannot be solved. For example, if the organic phase is absent ($s_o = 0$, $\omega_1^o = 0$), the organic species 1 equation (3.3) reduces to the identity $0 = 0$.

To handle this problem, one alternative is to solve a reduced set of equations in the region of two phase flow and couple this with the solution of the complete set in the three phase domain. This procedure would require that the migration of the boundary between the two regions be tracked in time. Coding would be extremely complex; different equations would have to be eliminated for different situations and different unknowns would have to be solved for in each region.

To surmount the difficulties described above, an alternative approach is employed in the model. At nodes where some fluid phase is absent, a small (negligible) saturation of that phase is prescribed in the initial conditions. The full system of mass balance equations may then be solved at every node in the domain. Numerical experimentation with this minimum saturation value, reveals that use of any value within the range 10^{-2} to 10^{-4} does not appreciably effect the rate of propagation of a front. Values below about 10^{-5} cause numerical difficulties in the band solver algorithm. For all the simulations described in this thesis, a negligible saturation value of 10^{-4} is employed. Phase saturations are not permitted to fall below this value.

To complete the set of initial conditions, the variables P_{wg}, P_{ow}, and ω_1^o must be prescribed at these negligible saturation nodes. In all the situations examined, it was found that the solution in the three phase region and the rate of front propagation were not sensitive to the choice of pressure in the negligible phase as long as this pressure was compatible with the small saturation value.

3.4 Evaluation of Coefficients

A glance at the coefficients tabulated in Appendix C.1 reveals that the mass balance equations contain many properties which must be evaluated in terms of the primary variables P_{ow}, P_{wg}, and ω_1^o. Table 3.1 contains a list of these parameters. Note that some parameters (specifically k_i, α_i, β_w, μ_w) are not functions of the primary variables over the pressure and composition ranges of interest. These parameters will, thus, not vary with time and are treated as constants in the computer model. The subscript i on k and α indicates that these properties may vary in space. Other parameters which appear in Table 3.1 are discussed in the following pages.

Dispersion/Diffusion Coefficients

In Chapter II Section 2.4, a general form of the dispersion tensor, $\underset{\approx}{D}^{\alpha}$, for an isotropic medium is introduced (equation (2.41)). The specific form of the dispersion tensor used in this model is a modification of the form first presented by Bear (1961). This form was also discussed by Scheidegger (1961) and later developed by Bear and Bachmat (1967) in their analysis of a capillary tube network:

$$D_{ij}^{\alpha} = D_{ij}^{m\alpha} + a_{ijkm}^{\alpha} \frac{\bar{v}_k^{\alpha} \bar{v}_m^{\alpha}}{\bar{v}^{\alpha}} \qquad (3.19a)$$

TABLE 3.1

List of Parameters

Parameters	Symbol	Dependence	Time Level
intrinsic permeability	k	x	c
compressibilites:			
matrix	α	x	c
water	β_w	constant	c
organic	β_o^P	P_{og}, ω_1^o	ℓ
	β_o^1	P_{og}, ω_1^o	ℓ
gas	β_g	ω_1^o	u
viscosities:	μ_w	constant	c
	μ_o	ω_1^o	ℓ
densities:	ρ^w	P_{wg}	ℓ
	ρ^o	P_{og}, ω_1^o	ℓ
	ρ^g	ω_1^o	u
saturations:	S_w	P_{wg}, P_{ow}, x	u
	S_o	P_{wg}, P_{ow}, x	u
	S_g	P_{wg}, P_{ow}, x	u
relative permeability:	k_{rw}	P_{wg}, P_{ow}, x	u
	k_{ro}	P_{wg}, P_{ow}, x	u
partition coefficients:	K_2^{wo}	$P_{wg}, P_{ow}, \omega_1^o$	u
	K_2^{gw}	$P_{wg}, P_{ow}, \omega_1^o$	u
dispersion/diffusion coefficients:	D^w	P_{wg}	ℓ
	D^o	P_{og}	ℓ
	D^g	x	c

x - indicates spatial dependence
ℓ - indicates parameter is lagged one time step
c - indicates constant in time
u - indicates parameter is updated after each interation

Here $\bar{v}^{\alpha}$ is the average fluid velocity, $\bar{v}_k^{\alpha}$ and $\bar{v}_m^{\alpha}$ are the velocity components in the coordinate directions, and $D_{ij}^{m\alpha}$ is the molecular diffusion tensor. a_{ijkm}^{α} is known as the dispersivity tensor. Equation (3.19a) was also obtained by Nikolaevskii (1959) using a statistical approach and analogy with the theory of turbulence. This form of the dispersion tensor has been used by many modelers in the tracing of solute migration (see, for example, Pickens and Grisak (1979), Pinder (1973), or Gray and Hoffman (1983b)).

In one space dimension, equation (3.19a) reduces to:

$$D^{\alpha} = D^{m\alpha} + a^{\alpha}\bar{v}^{\alpha} \qquad (3.19b)$$

The constant parameters $D^{m\alpha}$ and a^{α} must be input by the programmer. Fluid velocities are calculated internally by a finite difference discretization of Darcy's law (equation(2.25)). Dispersion coefficients are updated at the end of each time step. Recall that the dispersion/diffusion coefficient for the gas phase is equal to the molecular diffusive part, D^{mg}, of (3.19b).

Viscosity

The viscosities of liquids at low temperatures and moderate pressures are not found to be particularly affected by pressure changes (Reid, et al (1977)). Consequently, in this analysis, viscosities of the organic and water phases have been assumed independent of pressure.

The dependency of pure liquid viscosity on temperature is commonly expressed by Andrade's equation:

$$\mu_L = A\, e^{B/T} \tag{3.20}$$

where T is the temperature of the fluid and A and B are constants obtained from correlation with experimental data. This equation has been found to be generally valid in the temperature range from the freezing point to above the normal boiling point of the liquid (Reid, et al (1977)). Given a specific temperature, Andrade's equation may be used to calculate the viscosities of the two organic liquid phase components. These viscosities may then serve as input to the computer model.

Determination of the viscosity of a liquid mixture cannot generally be made to great accuracy if only the pure component properties are known. A simple correlation expression for the viscosity of a two-component liquid mixture is given by:

$$\mu_o = \mu_1^{x_1}\, \mu_2^{x_2} \tag{3.21}$$

where x_1 and x_2 are the mole fractions of components 1 and 2, respectively, in the mixture. This equation is commonly known as Arrhenius' equation and has been used by oil reservoir modelers in the absence of mixture viscosity data (Crookston, et al (1979)). Many other correlation equations have been proposed (see, for example, Reid, et al (1977) and Lohrenz, et al (1964)) which require some knowledge of the mixture's properties. The Arrhenius equation has been incorporated into this computer model, but could easily be replaced with a more sophisticated correlation if warranted by the available data.

Density

For modeling purposes, the liquid components of the system are assumed to fall under the classification of slightly compressible fluids. This classification encompasses most liquids treated in oil reservoir models (Crichlow (1977)) and also includes many other types of organic liquids (Lyman, et al (1982)). Slight compressibility implies that the compressibility of the liquid is small and may be regarded as constant over the pressure range of interest:

$$\beta_\alpha = \frac{1}{\rho^\alpha} \frac{\partial \rho^\alpha}{\partial P^\alpha} = \text{const} \tag{3.22}$$

Equation (3.22) may be integrated to yield an expression for liquid density:

$$\rho^\alpha = \rho^{\alpha b} \exp[\beta_\alpha (P^\alpha - P^{\alpha b})] \tag{3.23}$$

where $\rho^{\alpha b}$ is fluid density at the reference pressure $P^{\alpha b}$. An approximate expression for density may be derived by expanding the exponential in equation (3.23) and neglecting higher-order terms in β_α:

$$\rho^\alpha = \rho^{\alpha b}(1 + \beta_\alpha (P^\alpha - P^{\alpha b})) \tag{3.24}$$

Water phase density in the computer model is calculated directly from equation (3.24).

In the absence of mixture data, the density of the organic phase may be expressed by a weighted average of the densities of the individual liquid components (Crookston, et al (1979)):

$$\rho^{o} = \frac{1}{\sum_{i} \frac{\omega_i^{o}}{\rho^{oi}}} \tag{3.25}$$

where ρ^{oi} is the density of the liquid species i at the phase pressure P^{o}. ρ^{oi} is given by an equation of the form (3.23). Differentiation and manipulation of equation (3.25) yields the following expressions for organic phase compressibilities based on the properties of the phase components 1 and 2:

$$\beta_o^{1} = \frac{\rho^{o1} - \rho^{o2}}{\omega_1^{o}\,\rho^{o2} + \omega_2^{o}\,\rho^{o1}} \tag{3.26}$$

$$\beta_o^{P} = \beta_{o1} + \beta_{o2} - \frac{(\omega_2^{o}\,\beta_{o1}\rho^{o1} + \omega_1^{o}\,\beta_{o2}\rho^{o2})}{(\omega_1^{o}\,\rho^{o2} + \omega_2^{o}\,\rho^{o1})} \tag{3.27}$$

Relations (3.25), (3.26), and (3.27) along with expressions of the form (3.23) are used in the model to determine organic phase properties.

Gas phase density may be expressed by the gas law:

$$\rho^{g} = \frac{P^{g}\,M_g}{z\,R_u\,T} \tag{3.28}$$

where M_g is the molecular weight of the gas mixture, R_u is the universal gas constant and z is a compressibility factor which incorporates non-ideal behavior. Differentiation and manipulation of (3.28) yields the following expression for the compressibility of the gas phase (2.50):

$$\beta_g = \frac{M_2 - M_A}{M_2(1 - \omega_2^g) + M_A(\omega_2^g)} \tag{3.29}$$

where M_2 is the molecular weight of component 2 and M_A is the molecular weight of air. Equations (3.28) and (3.29) are used in the model to define gas phase properties.

Saturation and Permeability Relations

For application of the model, it is assumed that some empirical or analytical expressions that relate capillary pressures to saturations and relative permeabilities will be supplied to the program. These expressions may be obtained from laboratory work alone or from experimental work in conjunction with any one of a number of analytical predictive models (see, for example, Naar, et al (1962), Brooks and Corey (1966), or Mualem (1976)). Experimental work is normally performed on a two-phase system, and often only two-phase data is available to the modeler.

Leverett and Lewis (1941) were among the first to examine three-phase relative permeabilities in gas-oil-water systems. Under steady flow conditions in unconsolidated sands, they found relative permeability to water to be a function of water saturation alone. On the other hand, gas and oil relative permeabilities were found to depend on all saturations. Corey, et al (1956) conducted experiments on consolidated sandstones with similar results. In addition, they found that gas relative permeability was insensitive to the relative saturations of the two liquids. The type of functional dependence described above is a common premise of a large number of oil reservoir models (see, for example, Peery and Herron (1969) or Roebuck, et al (1969)).

Various methods for the extension of two-phase relative permeability data to the three-phase case have been proposed in the literature. Corey, et al (1956) offer an empirical equation for the calculation of water and oil relative permeabilities from measured gas permeabilities. Stone (1970) (1973) develops analytical models for the estimation of three-phase relative permeabilities from two-phase data using channel flow theory. Stone's models have been found to be in good agreement with experimental measurements and have been used extensively in oil reservoir models. Field verification, however, is still required (Aziz and Settari (1979)). Refinements of Stone's approach have been proposed by Dietrich and Bondor (1976) and Aziz and Settari (1979). Both of Stone's models have been employed in various portions of this thesis work.

A concept closely related to relative permeability which should be discussed at this point is that of a "residual saturation". Laboratory experiments such as those discussed above or the work of Schwille (1971) have revealed that there is a minimum saturation for a particular fluid below which that fluid cannot be made to flow. This residual saturation will vary with the type of fluid and the porous medium and must be determined in the laboratory. The residual saturation concept can be incorporated into a multiphase flow model by forcing the relative permeability of a given fluid to zero when its residual level is reached.

Channel flow theory, the basis of Stone's method, assumes that the porous medium may be described as a bundle of capillary tubes of variable cross section and that, in any tube, there will be at most only one mobile fluid. The wetting phase (water) will be located primarily in the small pore spaces and the nonwetting phase (gas) in the large pore spaces with the phase of intermediate wettability (organic) between them. In a three-phase system, water relative permeability and water-organic capillary pressure will be functions of water saturation alone and identical to the two-phase system relations. Similarly, gas phase relative permeability and gas-organic capillary pressure are the same functions of gas saturation in both three- and two-phase systems.

In Stone's 1970 model, the relative permeability of the organic phase (k_{ro}) is determined in the following manner. Normalized saturations (s_o^*, s_w^*, s_g^*) are defined. These saturations account for the presence of immobile fluids:

$$s_o^* = \frac{s_o - s_{om}}{1 - s_{wir} - s_{om}} \quad (\text{for } s_o \geq s_{om}) \qquad (3.30a)$$

$$s_w^* = \frac{s_w - s_{wir}}{1 - s_{wir} - s_{om}} \quad (\text{for } s_w \geq s_{wir}) \qquad (3.30b)$$

$$s_g^* = \frac{s_g}{1 - s_{wir} - s_{om}} \qquad (3.30c)$$

where s_{wir} is the residual water saturation and s_{om} is a minimum value of the residual organic saturation. Note that these definitions assume there is no residual gas saturation. The organic relative permeability is then given by the equation:

$$k_{ro} = s_o^* \Omega_w \Omega_g \tag{3.31}$$

where $\Omega_w = \frac{k_{row}}{1 - s_w^*}$ $\Omega_g = \frac{k_{rog}}{1 - s_g^*}$. k_{row} is the relative permeability of organic to water in a two-phase system and is a function of water saturation only. k_{rog} is the relative permeability of organic to gas in a two-phase system and is a function of gas saturation only. Equation (3.31) assumes impedance of organic flow by water and gas to be mutually independent events. In the region where $s_w \leq s_{wir}$, k_{ro} is assumed to be a unique function of gas saturation. For $s_o \leq s_{om}$, k_{ro} is set to zero.

An alternative predictive equation for organic relative permeability based on the same assumptions is given by Stone's 1973 method:

$$k_{ro} = (k_{row} + k_{rw})(k_{rog} + k_{rg}) - (k_{rw} + k_{rg}) \tag{3.32}$$

If equation (3.32) yields a negative value, k_{ro} is set to zero. Note that this second model does not require the specification of s_{om}.

The determination of saturations and relative permeabilities is further complicated by hysteresis effects. Sample wetting history has been found by many experimentalists (see, for example, Snell (1962) or Naar (1962)), to affect the saturation dependences of capillary pressures and relative permeabilities. No attempt was made in this thesis to accommodate cases of reversible wetting. All saturation and permeability curves used, however, were compatible with the directions of saturation change in the example problems.

Equilibrium Coefficients

The partition coefficients which are defined in equations (2.53) and appear in the equilibrium relations (2.54) must be input into the computer model in functional or tabular form. Functional dependency on all the primary variables is permitted.

3.5 Solution of the Nonlinear Matrix Equations

Equation (3.15) is a nonlinear matrix equation which must be solved for the $3 \times NN$ vector of unknowns $\underset{\sim}{u}$ at each time step. The technique which is used in the model to solve this system of equations is called the Newton-Raphson Method. This method is one of the best known iterative schemes and has been widely used for the solution of nonlinear equations. Many variations of this iteration technique exist (see, for example, the discussion in Ortega and Rheinboldt (1970)), but these variants will not be enumerated here. The standard or "variable tangent" Newton-Raphson method is the one employed in this model. Appendix B.3 contains a brief review of this technique.

The Newton-Raphson solution method may be simply represented as:

$$F_{ij}^{(\nu)} \delta_j^{(\nu)} = - f_i^{(\nu)} \qquad (3.33)$$

where $\delta_j^{(\nu)}$ is a correction vector defined by:

$$\delta_j^{(\nu)} = u_j^{(\nu+1)} - u_j^{(\nu)} \qquad (3.34)$$

Here (ν) represents the iteration level and F_{ij} is the Jacobian matrix of the vector function f_i. In this model, f_i is given by:

$$f_i(u_j^{n+1}) \equiv A_{ij} u_j^{n+1} - B_i \qquad (3.35)$$

A_{ij} and B_j are the coefficients which appear in equation (3.15) and are tabulated in Appendix C.1. The computer model solves the system of equations (3.15) by iterating on the system (3.33). Updated values of u_j are determined at each iteration level by equation (3.34). Note that both $\underset{\approx}{F}$ and $\underset{\sim}{f}$ must be recalculated at each iteration.

From examination of the structure of $\underset{\approx}{A}$ (see Figure 3.2), it can be seen that each f_i is dependent upon 9 distinct u_j at most. Thus, the matrix $\underset{\approx}{F}$ will have the same structure as that shown in the figure. The coefficients $(F_{kj})_i$ are too lengthy to be included here and have been tabulated in Appendix C.2.

Convergence of the iteration scheme depends to a great extent upon the quality of the initial guesses for u_j. In the model, values of the variables from the previous time step are used as the initial guesses, $u_j^{(0)}$. Experience indicates that, as long as the time step is kept within a certain range (dependent upon the boundary conditions and nodal spacing), these initial values are sufficiently close to the true solution to assure convergence of the scheme. In Chapter IV, convergence of the iteration method is demonstrated heuristically for some trial problems simulated by the computer model.

The procedure for evaluating the partial derivatives which appear in the coefficients F_{ij} is as follows. Let b be a parameter which is dependent on the variables u, v, w. An approximation to the partial derivative $\frac{\partial b}{\partial u}$ at (u_1, v_1, w_1) is given by the following relation:

$$\left.\frac{\partial b}{\partial u}\right|_1 \approx \frac{b(u_1 + \Delta u, v_1, w_1) - b(u_1, v_1, w_1)}{\Delta u} \qquad (3.36)$$

Note that in the limit as $\Delta u \to 0$, this approximation becomes an equality. Equation (3.36) is used in the model to calculate all derivatives. A centered approximation was not employed because of the additional computation cost it would entail. Three separate incremental values for the variables $(\Delta P_{ow}, \Delta P_{wg}, \Delta\omega_1^0)$ are input into the model. Since (3.36) is not a centered approximation, the sign of the increment as well as its magnitude may affect convergence of the Newton-Raphson scheme. In general, the sign of an increment should be consistent with the direction of change of that property over the time step. Experimentation with the magnitudes of the increments reveals that an upper limit must be placed on their values to ensure convergence of the iteration scheme. This upper limit is dependent upon the shape of the parameter curves.

* * * * * * * * *

A simplified flow chart of the computer model is given in Figure 3.3. After input data is read, parameters are evaluated through the functional relations discussed in Section 3.4. The matrix coefficients A_{ij} and the vector B_i are then assembled. To save computing time, $\underset{\approx}{A}$ and $\underset{\sim}{B}$ are split into two parts. One of the parts is made up of time invariant and lagged parameters and need not be reevaluated at every iteration. The right hand side vector $\underset{\sim}{f}$ is calculated from $\underset{\approx}{A}$, $\underset{\sim}{B}$ and $\underset{\sim}{u}$. Derivatives of parameters are computed next based on (3.36) and coefficients of the Newton-Raphson matrix are assembled. The matrix $\underset{\approx}{F}$ is placed in band storage form and a band solver is employed to compute the change in unknowns over the iteration. Use of the banded format reduces the storage requirements of the program significantly. The solver is specifically designed to handle non-symmetric banded matrices. It uses an upper triangularization and back substitution method of solution.

Once the solution is found, unknowns are updated and a convergence check is performed. The magnitude of the maximum change of a given unknown over the iteration is examined. If this magnitude does not exceed a certain limiting value, the solution is said to have converged in that variable.

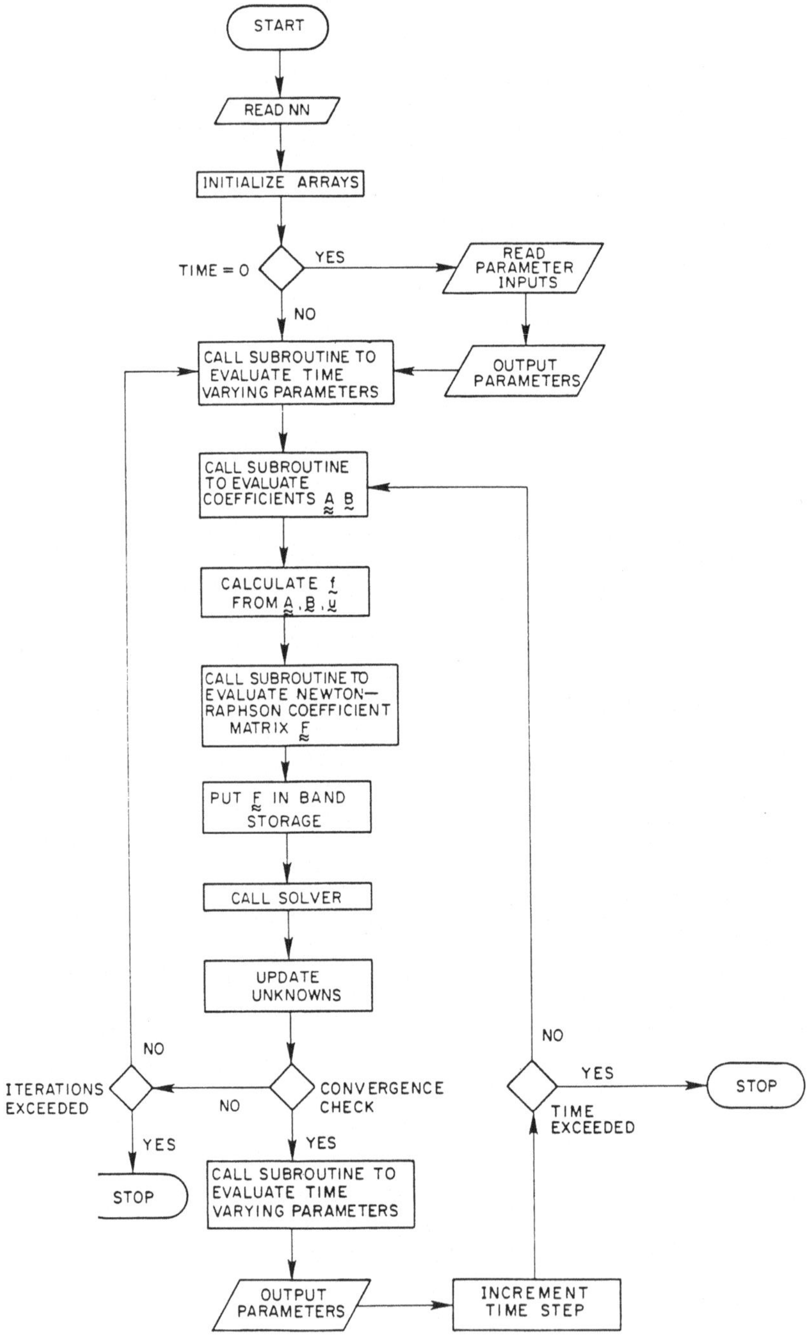

FIGURE 3.3 MODEL FLOW CHART

The convergence test limiting values are controlled by the programmer. If the method has not converged, parameters are updated, the non-constant portions of $\underset{\approx}{A}$ and $\underset{\approx}{B}$ are reevaluated, and the solution process is repeated. Execution will halt if either the specified maximum number of iterations or the simulation time is exceeded.

CHAPTER IV

COMPUTER SIMULATIONS IN ONE DIMENSION

This chapter presents some example simulations of multiphase flow using the computer model described in detail in Chapter III. Section 4.1 contains a series of oil contamination scenarios. Convergence of the solutions in space and time is explored, as is convergence of the iteration scheme. Mass balance calculations for the computer runs are also examined. Section 4.2 focuses on the movement of trichloroethylene (TCE) in a porous medium. Properties of TCE are introduced along with a detailed discussion of their theoretical and experimental basis. Simulations of TCE infiltration into a water saturated column are then presented.

4.1 Oil Contamination Simulations

A sketch of an example contamination scenario is given in Figure 4.1. Here a nonaqueous organic phase of known composition and constant head infiltrates into a homogeneous soil column which contains a residual saturation of water. Simulations were performed with the computer model for the case in which the organic phase was a heavy oil and the case in which the organic phase was a two-component mixture

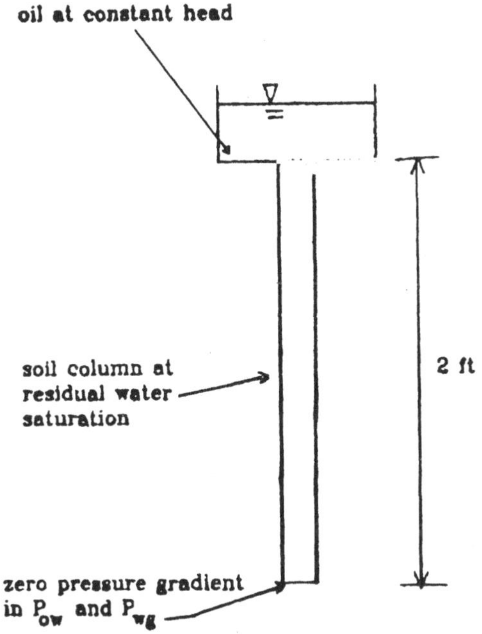

FIGURE 4.1: OIL CONTAMINATION SCHEMATIC - RUN 1

of heavy oil and propane. A listing of the model input parameters is given in Table 4.1. Most of these values were taken from an oil recovery simulation by Crookston, et al (1979). Note that dispersion/diffusion coefficients are constant in time for these simulations.

Saturations and relative permeabilities for the simulations were calculated from the following functional relations (Coats (1980b)):

$$s_w = 0.5 - 4.63 \times 10^{-3} P_{ow} \tag{4.1}$$

$$s_o = 8.21 \times 10^{-4} P_{og} + 6.45 \times 10^{-1} - s_w \tag{4.2}$$

$$\begin{cases} K_{rw} = K_{rwro} \left(\dfrac{s_w - s_{wir}}{1 - s_{orw} - s_{wir}}\right)^{n_w} & \text{for } s_w > s_{wir} \\ K_{rw} = 0 \quad \text{for } s_w \leq s_{wir} \end{cases} \tag{4.3}$$

$$\begin{cases} K_{row} = K_{rorw} \left(\dfrac{1 - s_{orw} - s_w}{1 - s_{orw} - s_{wir}}\right)^{n_{ow}} & \text{for } s_w < 1 - s_{orw} \\ K_{row} = 0 \quad \text{for } s_w \geq 1 - s_{orw} \end{cases} \tag{4.4}$$

$$\begin{cases} K_{rog} = K_{rorw} \left(\dfrac{1 - s_{wir} - s_{org} - s_g}{1 - s_{wir} - s_{org}}\right)^{n_{og}} & \text{for } s_g < 1 - s_{wir} - s_{org} \\ K_{rog} = 0 \quad \text{for } s_g \geq 1 - s_{wir} - s_{org} \end{cases} \tag{4.5}$$

TABLE 4.1

Parameters Used in Oil Contamination Simulations

Parameter	Value	Units	Reference Equation
water:			
M_w	18.02		
μ_w	2.735×10^{-5}	lb·sec/ft^2	
β_w	2.78×10^{-8}	psf^{-1}	(3.22)
ρ^{wb}, p^{wb}	64.6, 1.70×10^4	lb/ft^3, psf	(3.24)
heavy oil (component 1):			
M_1	170.0		
μ_1	4.71×10^{-4}	lb·sec/ft^2	(3.21)
β_{o1}	6.94×10^{-8}	psf^{-1}	(3.23)
ρ^{o1b}, p^{o1b}	45.6, 2.12×10^3	lb/ft^3, psf	(3.23)
propane (component 2):			
M_2	44.09		
μ_2	2.767×10^{-6}	lb·sec/ft^2	(3.21)
β_{o2}	1.54×10^{-6}	psf^{-1}	(3.23)
ρ^{o2b}, p^{o2b}	29.98, 1.44×10^5	b/ft^3, psf	(3.23)
soil matrix:			
ε_i	0.30		
α	0.0	psf^{-1}	(2.14)
k	10^{-8}	ft^2	(2.25)

Table 4.1 (cont)

Parameter	Value	Units	Reference Equation
gas phase:			
M_A	28.97		(3.29)
Z, P^g	1.0, 2.12 x 10^3	- , psf	(3.28)
g	32.2	ft/sec^2	(2.25)
T	15°C = 518.67°R		(3.28)
permeability:			(4.3)-(4.6)
k_{rwro}	0.70		
k_{rorw}	0.90		
k_{rgro}	0.90		
s_{wir}	0.20		
s_{orw}	0.25		
s_{org}	0.10		
s_{gr}	0.10		
n_w	2.0		
n_{ow}	2.0		
n_{og}	2.0		
n_g	2.0		
dispersion/diffusion:			
D^w	6.0 x 10^{-4}	ft^2/sec	constant values used
D^o	3.0 x 10^{-4}	ft^2/sec	constant values used
D^g	1.08 x 10^{-3}	ft^2/sec	constant values used
partition coefficients (molar):			
K_p^g	0.1		(4.7)
K_p^w	0.01		(4.7)

$$\begin{cases} K_{rg} = K_{rgro}\left(\dfrac{s_g - s_{gr}}{1 - s_{wir} - s_{org} - s_{gr}}\right)^{n_g} & \text{for } s_g > s_{gr} \\ K_{rg} = 0 \quad \text{for } s_g \leq s_{gr} \end{cases} \tag{4.6}$$

Here K_{rwro}, K_{rorw}, K_{rgro} are the two-phase relative permeabilities of water to residual organic, organic to residual water, and gas to residual organic, respectively. Residual saturation values for water, organic to water, organic to gas, and gas are given by the parameters s_{wir}, s_{orw}, s_{org}, and s_{gr}. n_w, n_{ow}, n_{og}, and n_g are experimental constants. A list of the values of these parameters used in the simulations is included in Table 4.1.

Note that the pressure-saturation relations (4.1) and (4.2) are linear. Thus, derivatives of saturations with respect to capillary pressures will be constant in time and need not be reevaluated at every iteration level. Also note that the relative permeabilities calculated by (4.3) - (4.6) are for two phase systems. Three phase permeabilities are estimated in the model by Stone's second method (equation (3.32)).

Mass partition equations were assumed to have the following functional form (after Crookston, <u>et al</u> (1979)):

$$\begin{aligned} x_2^g &= K_p^g\left[\frac{s_o}{s_o + 10^{-4}}\right] x_2^o \\ x_2^w &= K_p^w\left[\frac{s_o}{s_o + 10^{-4}}\right] x_2^o \end{aligned} \tag{4.7}$$

where x_2^{α} is the mole fraction of component 2 in the α phase and K_p^g and K_p^w are temperature dependent molar partition coefficients. K_p^g and K_p^w are input directly to the computer model, and then these values, along with the relations (4.7), are used to calculate mole fractions. ω_2^w, ω_2^g, K_2^{gw}, and K_2^{wo} are then computed by the use of simple mole-to-mass conversion relations.

RUN 1

The results of an example computer simulation are shown in Figure 4.2. Here a heavy oil is propagating downwards through a vertical soil column under a constant head. Initial and boundary conditions along with additional parameters required for this simulation are given in Table 4.2. For this example, the spatial domain was divided into eight equal intervals and a time step of 150 seconds was employed. Numbers labeling the curves in Figure 4.2 refer to the number of time steps required to attain a given saturation profile. A total time of 3000 seconds was simulated. The small triangles on the curves are the actual computed values. For aesthetic reasons, smooth curves have been fit through these points in the figure. Note that the maximum oil saturation attainable is 0.70, due to the presence of residual water and air in the soil column. Also note that there is no interphase mass exchange involved in this example.

For the specific convergence criteria parameters listed in Table 4.2, the model averaged 6 iterations per time step. Results were generated by an IBM 3081 system in conjunction with the FORTRAN G compiler.

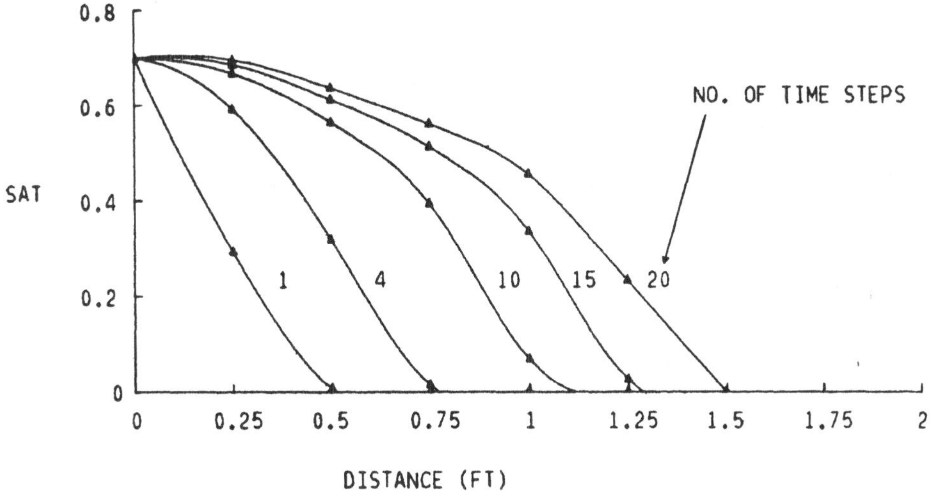

FIGURE 4.2: PROPAGATION OF OIL PHASE - RUN 1

TABLE 4.2

Boundary Conditions and Parameter Inputs for RUN #1

Initial Conditions

$$\left.\begin{aligned} s_w &= 0.2 \\ s_o &\simeq 0 \end{aligned}\right\} \Longrightarrow \begin{aligned} P_{ow} &= 64.8 \text{ psf} \\ P_{wg} &= -606.0 \text{ psf} \end{aligned} \quad \text{everywhere}$$

Boundary Conditions

at top: $\omega_1^o = 1.0$

$P_{ow} = 64.8$ psf

$P_{wg} = 307.0$ psf

at bottom: $\dfrac{\partial \omega_1^o}{\partial x} = 0$

$\dfrac{\partial P_{ow}}{\partial x} = 0$

$\dfrac{\partial P_{wg}}{\partial x} = 0$

Convergence Criteria

P_{ow}: 1.0×10^{-2}

P_{wg}: 1.0×10^{-2}

ω_1^o : 1.0×10^{-8}

Time

Max time: 3.0×10^3 sec.

Time step: 1.5×10^2 sec.

A total of 2.02 seconds of CPU time was required to simulate 1500 seconds of model time, including parameter output at each time step. This breaks down to approximately 0.20 seconds of CPU time per time step or roughly 0.033 seconds per iteration. Total core usage consisted of approximately 130 K for the object code and machine overhead and (2.27 x NN) + 5 K for array storage. The program was run in double precision.

The convergence characteristics for the iteration scheme are illustrated by Figures 4.3(a) and (b). Figure 4.3(a) is a semi-log plot of the absolute value of the maximum residual versus the number of iterations for a representative time step. Here residual, R_i^m, is defined as the quantity:

$$R_i^m \equiv A_{ij} u_j^m - B_i \tag{4.8}$$

Recall that $\left|R_i^m\right|_{max}$ tends towards zero as the exact solution to the system of nonlinear equations is approached (see equation (3.15)). Note that the residual decreases almost ten orders of magnitude in Figure 4.3(a). The only observed limit to this decrease was that imposed by computer storage of significant figures.

A relationship between successive residuals is suggested by the log-log graph in Figure 4.3(b). Here the maximum residual at iteration level $m+1$ is plotted against the corresponding residual at iteration level m. The points lie approximately along a straight line fit to the data by a least squares method. Utilizing the slope and intercept

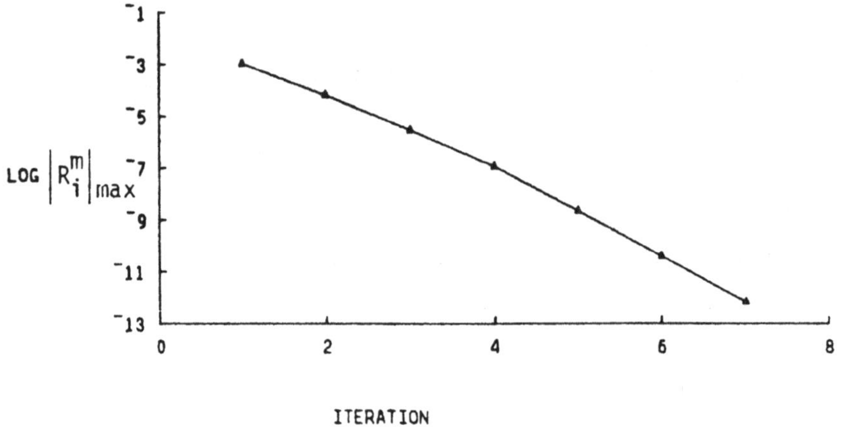

FIGURE 4.3(A): CONVERGENCE OF ITERATION SCHEME

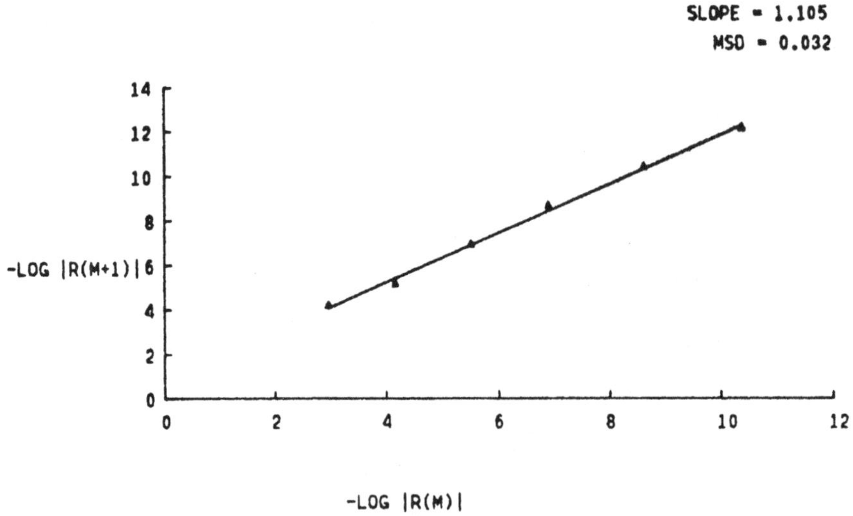

FIGURE 4.3(B): CONVERGENCE OF ITERATION SCHEME

of this line, the following equation for the residual may be formulated:

$$|R^{m+1}|_{max} = 0.16(|R^m|_{max})^{1.105} \tag{4.9}$$

It must also be mentioned here that the computed solution to the mass balance equations was found to be relatively insensitive to the chosen convergence criteria. As long as this criteria was less than about 1% of a variable's magnitude, little change was observed in the solution when this criteria was reduced. For example, when the convergence criteria was lowered by three orders of magnitude, the solution changed only in the fifth significant figure.

Figure 4.4(a) illustrates the effect of mesh refinement on the oil saturation profile. In the figure, solutions for various nodal spacings are compared at a fixed time and time step. Note that, as the spacing becomes smaller, the saturation front steepens. It is also apparent that the change in saturation profile decreases with each successive refinement of the mesh. As Δx diminishes, successive solutions become virtually indistinguishable. The rate of convergence with respect to Δx is examined in Figure 4.4(b). Here solutions at various nodal spacings are compared to the solution at a very small spacing ($\Delta x = 0.0833$). This solution is assumed to be close to the true solution and all residuals are expressed in terms of it. Note that as Δx decreases, the log-log plot becomes a straight line whose slope is 3.27. If a least squares line is fit to all four points in the plot, however, the slope of this fitted line is found to be 2.2. A spatial order

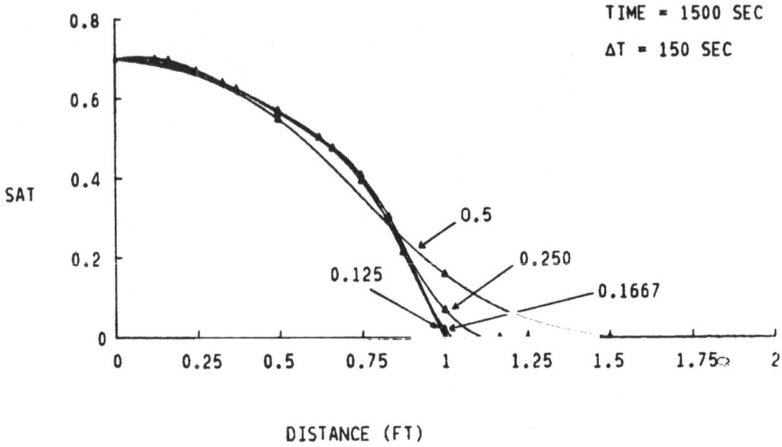

FIGURE 4.4(A): NODAL SPACING COMPARISON

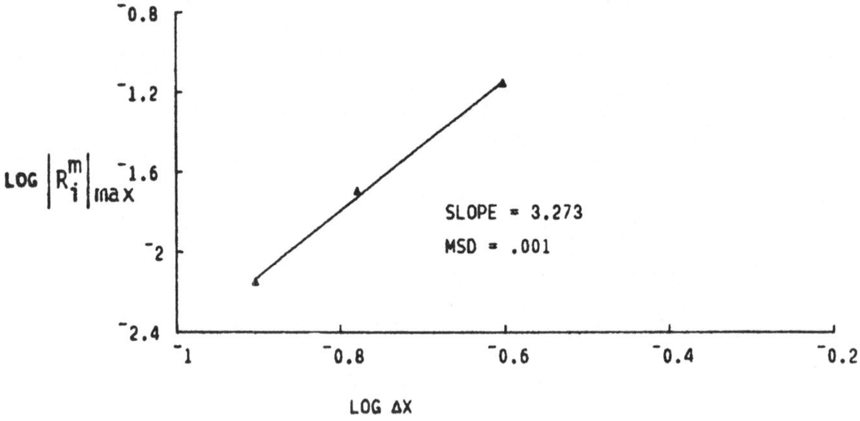

FIGURE 4.4(B): SPATIAL CONVERGENCE

of convergence of 2.0 was anticipated for a simulation with equal nodal spacing (see Section 3.2 and Appendix B.2).

Figure 4.5(a) examines the effect of a decreasing time step on the saturation profile. Solutions for various time step sizes are compared at a specific time for fixed convergence criteria and nodal spacing. As with the nodal spacing comparison above, the saturation curve is found to steepen with successive refinements and solutions become indistinguishable on the plot as Δt decreases. In Figure 4.5(b), the rate of convergence in time is investigated. Here residuals are calculated with respect to the solution for a very small Δt (37.5), assumed to be close to the true solution. On the log-log plot, data points are fit by a least mean squares line. The order of convergence of the scheme in time is, thus, shown to be roughly 0.982. This result is close to that of first-order convergence which would be anticipated from the use of a backward differencing scheme.

Calculations were performed to determine whether mass was conserved in Run #1. Consider Figure 4.6(a). It depicts the finite difference mesh used for the run. The total mass of oil in the system at any given time may be approximated by the area under the rectangles shown in the figure. Mathematically speaking, this approximation is equivalent to the following expression:

$$\text{mass}^n = \sum_i (\rho^o \, \varepsilon \, s_o \, \Delta x)_i^n \tag{4.10}$$

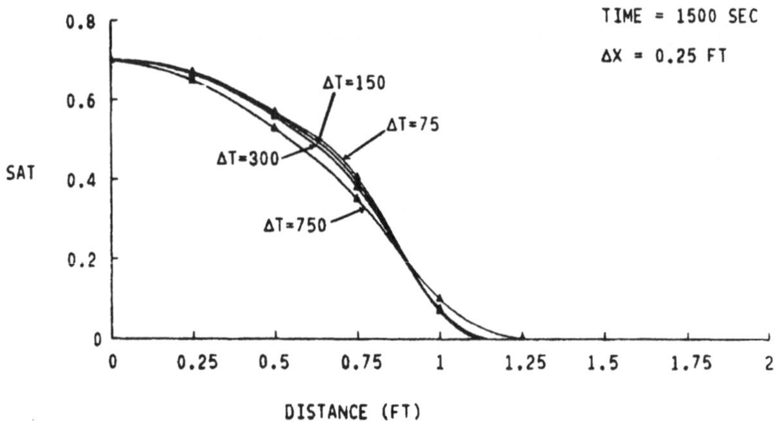

FIGURE 4.5(A): TIME STEP COMPARISON

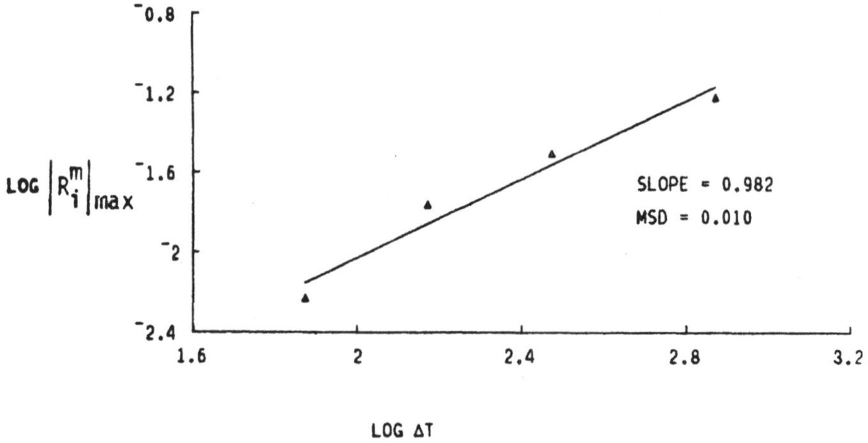

FIGURE 4.5(B): CONVERGENCE IN TIME

Note that, at the end nodes, half of the element length should be used for Δx. The change in oil mass over a time step may then be expressed simply as:

$$\Delta \text{mass} = \text{mass}^{n+1} - \text{mass}^{n} \qquad (4.11)$$

If the solution scheme conserves mass, the change in mass over a time step should be equal to the flux which enters the system during that time interval. Many methods for estimating this flux are possible. Note that the slashed area at node 1 in Figure 4.6(a) remains constant during the simulation due to the constant pressure conditions at node 1. Thus, the inclusion or exclusion of this rectangle in the mass calculation will have no effect on Δmass. Also note that fluxes into and out of this rectangle should be equal. The pressure gradient at node 1 could be computed by a first-order, one-sided, difference expression. If the flux at the point $1+\frac{1}{2}$ is considered, however, a more accurate, centered, expression may be written for the pressure gradient. The flux at the point $1+\frac{1}{2}$ may be written as:

$$\text{flux} = (\rho^{o}\tau_{o})_{1+\frac{1}{2}} \left[\frac{P_2^o - P_1^o}{\Delta x_1} - \gamma_{1+\frac{1}{2}}^{o}\right] \Delta t \qquad (4.12)$$

An average flux over the interval Δt may be computed from (4.12) using parameters evaluated at time level n+1. A measure of the mass conservation properties of the model may then be expressed as:

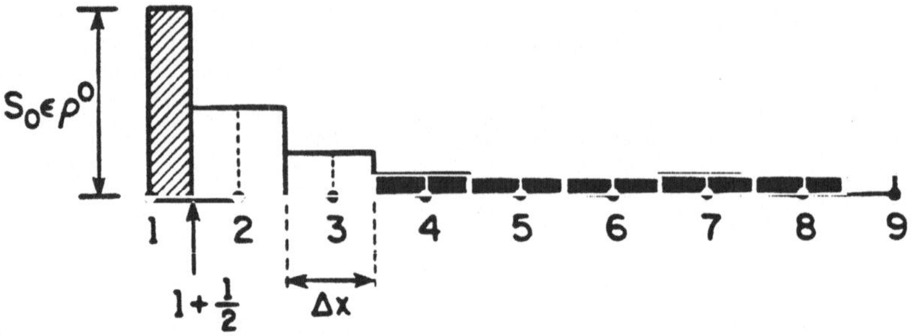

FIGURE 4.6 (a): MASS CALCULATION

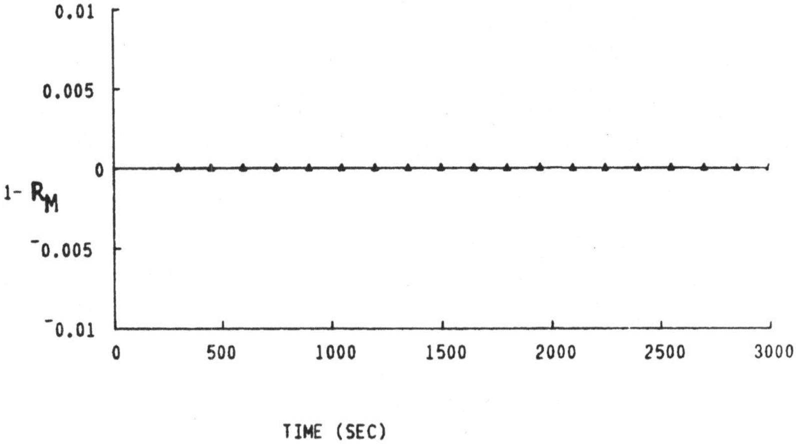

FIGURE 4.6(B): CONSERVATION OF MASS FOR RUN 1

$$R_m = \frac{\Delta \text{mass}}{\text{flux}_{ave}} \tag{4.13}$$

As noted previously, R_m will be equal to 1 if mass is conserved.

The deviation of R_m from unity is plotted in Figure 4.6(b). Although it cannot be distinguished in the figure, this deviation never exceeds 0.00005. Because of the way in which the boundary conditions are imposed on the system, there is an abrupt change in mass at node 1 at the time 0^+. For this reason, mass^1 does not truly reflect the change of mass over the first time step and results of the mass balance calculation for this time step are not meaningful. Note that the R_m value at time step 1 has been omitted from the plot. No improvement in the mass balance figures was observed for smaller Δx or Δt increments.

The effect of upstream weighting on the solution is illustrated by Figure 4.7(a). Here, full upstream weighting was used on the oil phase ($\theta_o = 1$ in equation (3.8)). Two saturation profiles are shown in the figure for a specific time, Δx, and Δt. Note that the upstream weighted profile is shifted downstream from the centered curve. A mass balance calculation on the upstream weighted computer run indicates that, for this particular problem, mass conservation is not as good as with the centered approximation (see Figure 4.7(b)). Thus, the centered solution is the preferred one.

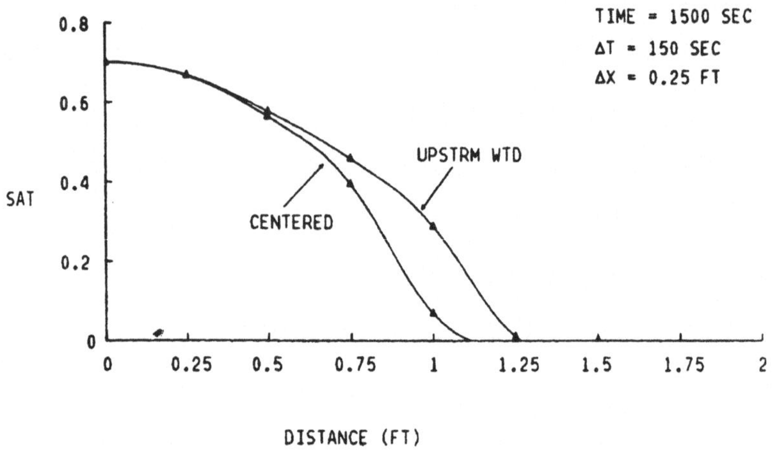

FIGURE 4.7(A): EFFECT OF UPSTREAM WEIGHTING

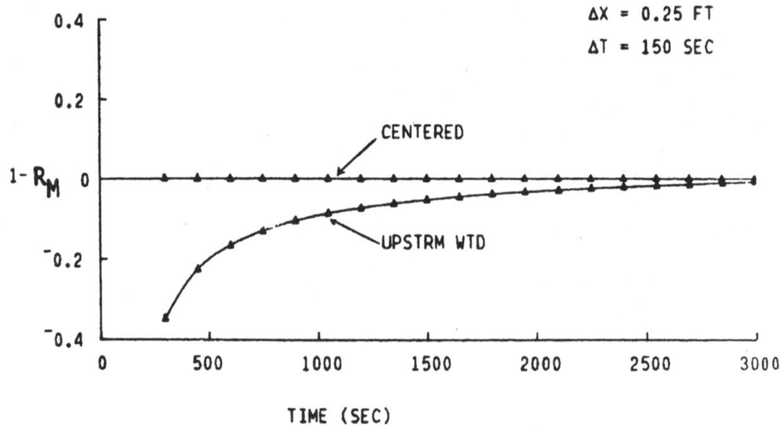

FIGURE 4.7(B): MASS CONSERVATION COMPARISON

RUN 2

A second test of the one-dimensional model was made on the flow scenario depicted in Figure 4.8. Here a horizontal, homogeneous soil column is almost completely saturated with the oil phase and there is no flux of fluids. A small, fixed concentration of a second component is introduced at the left-most boundary and a zero concentration is maintained at the other end of the domain. All partition coefficients are set to zero. Since pressures are constant and gravity terms are zero, the governing equation for the mass balance of component 2 (equation (3.4)) simplifies to the following differential equation:

$$\rho^o \varepsilon\, s_o \frac{\partial \omega_2^o}{\partial t} + \omega_2^o s_o \varepsilon\, \beta_o^1 \frac{\partial \omega_1^o}{\partial t} - \frac{\partial}{\partial x}\left(\rho^o \varepsilon\, s_o D^o \frac{\partial \omega_2^o}{\partial x}\right) = 0 \qquad (4.14)$$

If component 2 has the same mass density as component 1, then $\beta_o^1 = 0$ and $\rho^o = \text{const}$, and for constant D^o, equation (4.14) reduces to the familiar diffusion equation:

$$D^o \frac{\partial^2 \omega_2^o}{\partial x^2} = \frac{\partial \omega_2^o}{\partial t} \qquad (4.15)$$

The analytical solution of this equation for the boundary conditions indicated in Figure 4.8 and the initial condition $\omega_2^o(x,0) = 0.0$ may be written as (Powers (1972)):

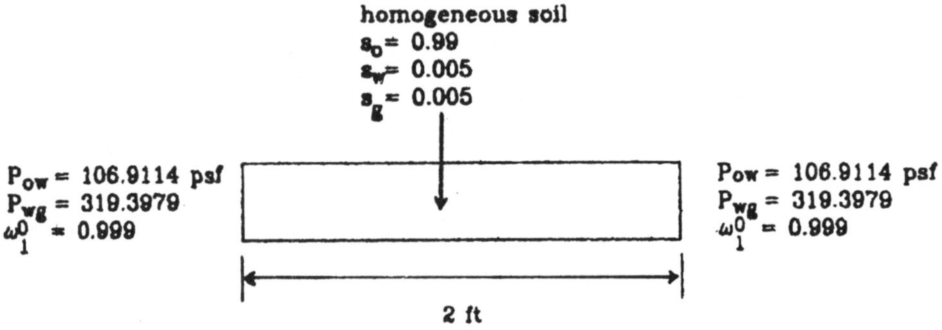

FIGURE 4.8: DIFFUSION IN THE OIL PHASE – RUN 2

$$\omega_2^o(x,t) = T(x,t) + 0.001 + (-0.001)\,\frac{x}{L} \tag{4.16}$$

where

$$T(x,t) = \frac{-2}{\pi} \sum_{1}^{\infty} \frac{0.001}{n} \sin(\lambda_n x) \exp(-\lambda_n^2 D^o t)$$

and

$$\lambda_n = \frac{n\pi}{L} .$$

For simulation run #2, $D^o = 3 \times 10^{-4}$ and $L = 2.0$.

Figure 4.9(a) shows a plot of this analytical solution at a time $t = 1000$ sec. The other curves shown in the figure are computer model solutions of the full system of mass balance equations (3.15) for various space and time increments. Note that as these increments are decreased, the analytical solution is approached. Figure 4.9(b) is a plot of the steady-state solution. Note that the model solution at large time exactly overlies this linear solution. It must also be mentioned that no degradation of the model steady-state solution was observed with time.

RUN 3

One final contamination scenario will be discussed here to illustrate the full capabilities of the multiphase model. Again, as in Figure 4.1, an oil phase of known composition infiltrates into a vertical soil column at residual water saturation. The initial and boundary conditions for Run 3 are listed in Table 4.3. In this simulation, the infiltrating oil phase is composed of both heavy oil and

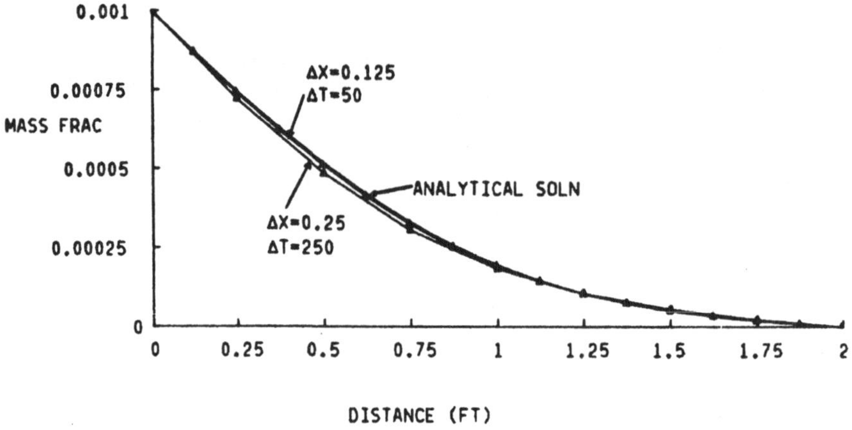

FIGURE 4.9(A): SOLUTION TO DIFFUSION PROBLEM AT T=1000 SEC

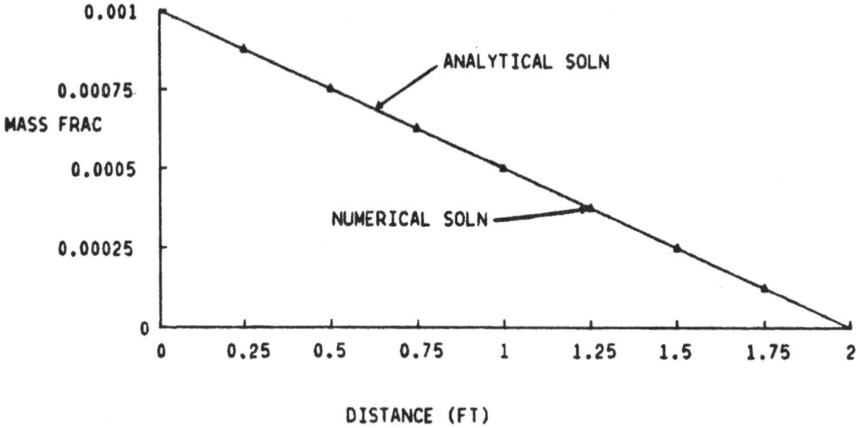

FIGURE 4.9(B): STEADY STATE SOLUTION TO DIFFUSION PROBLEM

TABLE 4.3

Boundary and Initial Conditions for RUN #3

Initial Conditions

$$\left.\begin{array}{l} s_w = 0.2 \\ s_o \simeq 0 \end{array}\right\} \quad \begin{array}{l} P_{ow} = 64.8 \text{ psf} \\ P_{wg} = -606.0 \text{ psf} \end{array} \quad \text{everywhere}$$

$$\omega_1^o = 1.0$$

Boundary Conditions

at top:

$$\omega_1^o = 0.999$$

$$P_{ow} = 64.8 \text{ psf}$$

$$P_{wg} = 307.0 \text{ psf}$$

at bottom:

$$\omega_1^o = 1.0$$

$$\frac{\partial P_{ow}}{\partial x} = 0.0$$

$$\frac{\partial P_{wg}}{\partial x} = 0.0$$

propane. As the oil phase move downwards, the propane may volatize or dissolve in the water phase.

Some simulation results for various times are plotted in Figures 4.10(a), (b), and (c). Figure 4.10(a) shows the progress of the infiltrating oil phase. The bounds of this infiltration should be compared to the movement of the contaminant in the gas and water phases (Figures 4.10(b) and (c)). Note that a contaminated plume of gas extends beyond the main oil body. The movement of the contaminant in the water and gas phases is controlled solely by diffusion since there is no convective movement of these phases. Gas diffusion in this problem is about an order of magnitude greater than diffusion in the water phase, and thus, this is the primary mechanism for contaminant transport. The presence of an accompanying plume in the water phase is due, for the most part, to the local equilibrium assumption, i.e. the concentration in the water phase must be in equilibrium with the gas phase concentration.

4.2 TCE Simulations

This section of Chapter IV focuses on the application of the 1-D model to a different fluid system: trichloroethylene, water, and air. Trichloroethylene (TCE) is a degreasing agent commonly used by industry and in private households. Due to its widespread manufacture and use, TCE is often detected as a contaminant of groundwater supplies. It is representative of a large group of chlorinated hydrocarbons with similar characteristics. Many of the physical properties of TCE have been

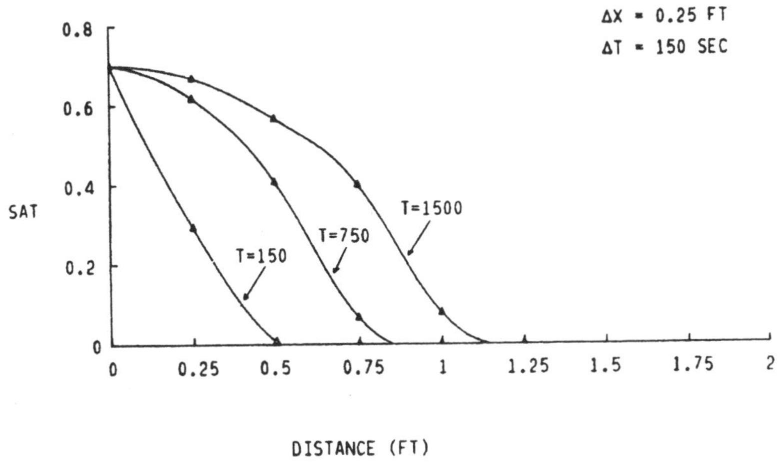

FIGURE 4.10(A): PROPAGATION OF OIL PHASE - RUN 3

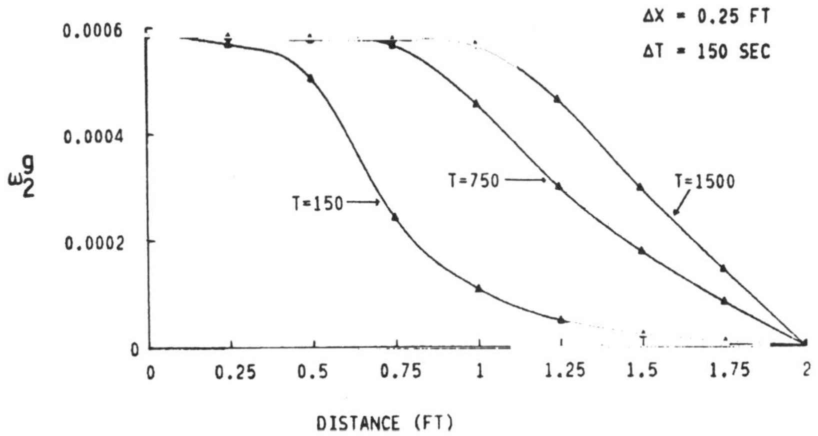

FIGURE 4.10(B): CONTAMINANT IN GAS PHASE - RUN 3

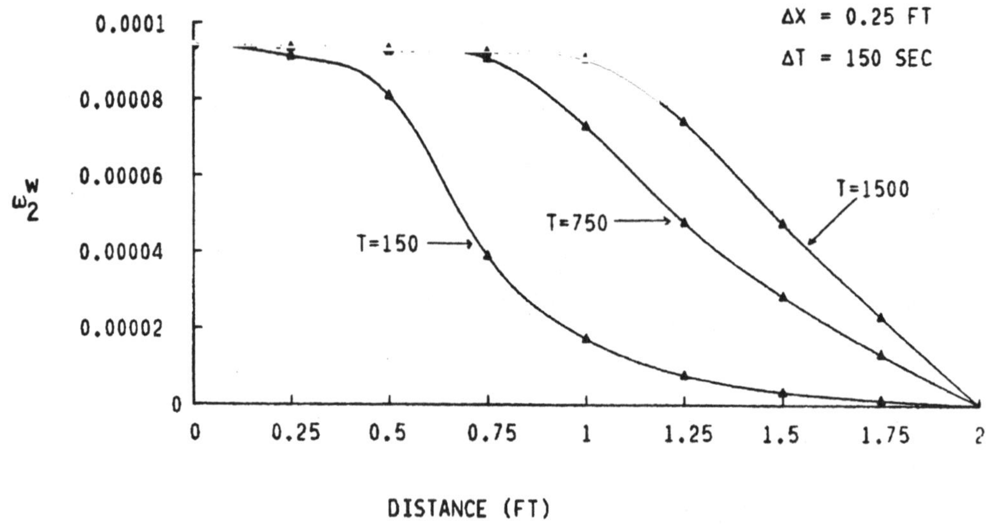

FIGURE 4.10(C): CONTAMINANT IN WATER PHASE - RUN 3

investigated and are accessible in the literature (see, for example, Gallant (1966) or Lyman, et al (1982)). This compound is a liquid at room temperature and is both volatile and slightly water soluble. A tabulation of the TCE parameters used in the computer simulations is given in Table 4.4. Note that the compressibility of TCE was not available and has been assumed, in the absence of data, to be negligible. Also listed in Table 4.4 are the pertinent properties of water and air in cgs units.

Two-phase saturation and relative permeability inputs for the computer simulations were derived, in part, from the experimental work of Lin, et al (1982) with the fluids TCE and water. They made laboratory measurements of pressure-saturation relations for water-air and TCE-air systems in homogeneous sand columns. A listing of the properties of the soil matrix used in these experiments is also included in Table 4.4. Much of the laboratory work is described in the 1982 report mentioned above, but in addition, some of the data used to develop the curves in this thesis were obtained through private communications with Dr. Lin.

Using a nonlinear least-squares software package developed by van Genuchten (1978), Lin, et al (1982) fit the experimental data to a pressure-saturation equation of the form:

$$s_{\alpha} = s_{\alpha r} + \frac{s_{\alpha s} - s_{\alpha r}}{[1 + (ah)^n]^m} \tag{4.17}$$

Here, $h = \frac{-P_c}{\rho^{\alpha} g}$ and $m = 1 - 1/n$. n and a are the parameters to be fit. The subscripts s and r indicate saturated and residual levels of

TABLE 4.4

Parameters Used in TCE 1-D Simulations

Parameter	Value	Units	Reference Equation
Water:			
M_w	18.02		
μ_w	1.0019×10^{-2}	poise	
β_w	4.531×10^{-11}	$cm^2/dyne$	(3.22)
ρ^{wb}, P^{wb}	0.9997964, 1.0133×10^6	g/cm^3, $dyne/cm^2$	(3.22)
TCE:			
M_1	131.4		
μ_1	5.8×10^{-1}	poise	(3.21)
β_{o1}	~ 0.0		(3.23)
ρ^{o1b}, P^{o1b}	1.4657, 1.0133×10^6	g/cm^3, $dyne/cm^2$	(3.23)
Air:			
M_A	28.97		(3.29)
Z, P^g	1.0, 1.0133×10^6	-, $dyne/cm^2$	(3.28)
g	9.80665×10^2	cm/sec^2	(2.25)
T	293.15	°K	(3.28)
Matrix:			
ε_i	0.36		
α	2.0×10^{-10}	$cm^2/dyne$	(2.14)
k	5.8231×10^{-7}	cm^2	(2.25)

Table 4.4 (cont)

Parameter	Value	Units	Reference Equation
Residual Saturations:			
s_{wir}	0.306		(3.30)
s_{om}	0.17		(3.30)
Dispersion/ Diffusion:			
D^{mg}	0.039	cm^2/sec	(3.19b)
D^{mw}, a^w	8.434×10^{-6}, 0.1	cm^2/sec, cm	(3.19b),(4.26)
D^{mo}, a^o	0.0, 0.0	cm^2/sec, cm	(3.19b)
Partition Coefficients:			
K_p^w	3.018×10^{-4}	-	(4.28)
K_p^g	5.549×10^2	-	(4.30)
Convergence Criteria:			
ΔP_{ow}	10.0	dyne/cm^2	
ΔP_{wg}	10.0	dyne/cm^2	
$\Delta \omega_1^o$	0.001	-	

the phase α . P_c is the capillary pressure between the wetting and nonwetting fluids. Equation (4.17) has been used to model soil moisture retention data in many studies (see, for example, Ahuja and Schwartzendruber (1972) or Haverkamp, et al (1977)). Fitted pressure-saturation curves are plotted in Figures 4.11(a) and 4.12(a) for the cases of TCE imbibition and water drainage respectively. These relations have been input directly to the computer model.

Using the theory of Mualem (1976), expressions for two-phase relative permeabilities may be derived from pressure-saturation relations. Mualem's permeability model is based on a probabilistic approach in which a porous matrix is conceived of as a homogeneous medium with interconnected pores of varying radii. In his derivation, Mualem replaces the pore configurations by capillary elements and assumes that tortuosity and correlation factors are power functions of moisture content. His predictive equation for relative permeability is given by:

$$k_{r\alpha} = s_{\alpha e}^{\frac{1}{2}} \left[\int_0^{s_{\alpha e}} \frac{1}{h(s)}\, ds \Big/ \int_0^1 \frac{1}{h(s)}\, ds \right]^2 \qquad (4.18)$$

where $s_{\alpha e} = \dfrac{s_\alpha - s_{\alpha r}}{s_{\alpha s} - s_{\alpha r}}$ is called the effective saturation. Using the pressure-saturation relation (4.17), van Genuchten (1978) produces the following closed form solution of equation (4.18):

$$k_{r\alpha} = \frac{[1 - (ah)^{n-1}\,(1 + (ah)^n)^{-m}]^2}{[1 + (ah)^n]^{m/2}} \qquad (4.19)$$

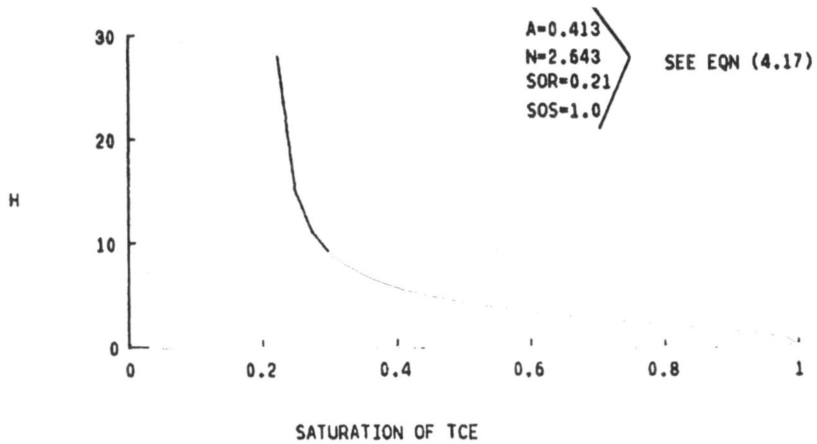

FIGURE 4.11(A): CAPILLARY CURVE FOR TCE-GAS SYSTEM (IMBIBITION)

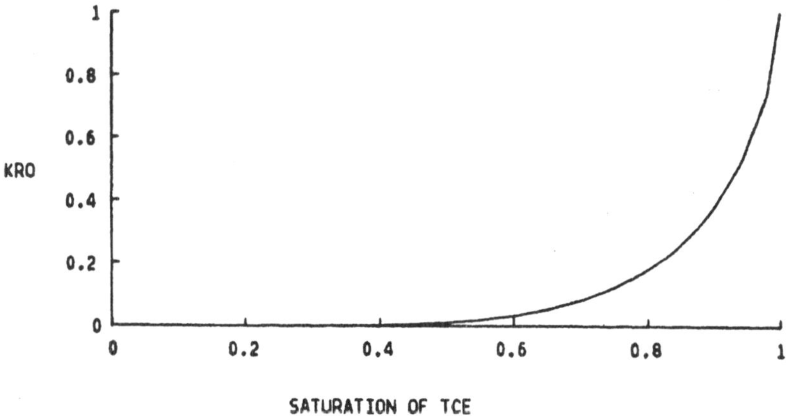

FIGURE 4.11(B): RELATIVE PERMEABILITY OF TCE IN TCE-GAS SYSTEM

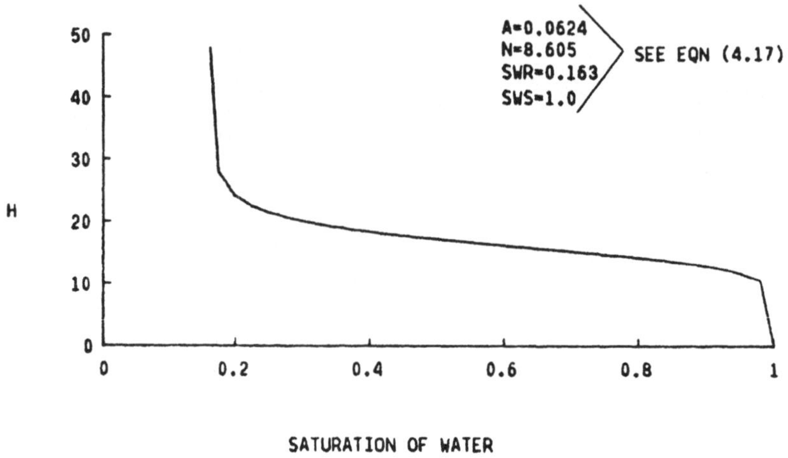

FIGURE 4.12(A): CAPILLARY CURVE FOR WATER-GAS SYSTEM (DRAINAGE)

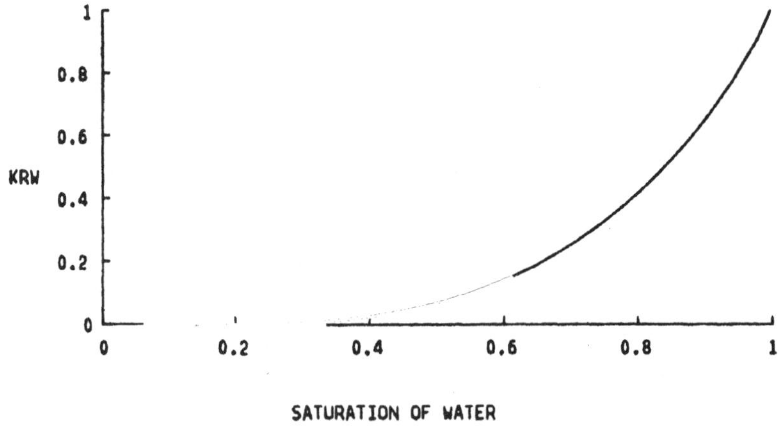

FIGURE 4.12(B): RELATIVE PERMEABILITY OF WATER IN WATER-GAS SYSTEM

Figures 4.11(b) and 4.12(b) show the predicted relative permeability curves for the TCE-air and water-air systems based on equation (4.19). Due to the absence of permeability data, these curves have been incorporated into the model to calculate two-phase relative permeabilities.

Equation (4.17) was also fit to TCE-water saturation data (Lin (1982)) and the result is plotted in Figure 4.13(a). The curve shown is for the case of TCE imbibition. For this two-fluid system, direct measurements of relative permeability were also available (Lin, et al (1982)). These permeability data points are plotted in Figure 4.13(b) along with two cubic polynomial curves which were fit to the data by an IBM curvilinear regression package, RLFOR. The polynomial curves have the form:

$$k_{r\alpha} = b_0 + b_1 s_w + b_2 s_w^2 + b_3 s_w^3 \qquad (4.20)$$

where b_i, $i=0,3$ are constants. The values of these coefficients for each curve are given in the figure. Observe that one data point was discarded in this analysis. By definition, the relative permeability of water in a saturated water system is equal to unity. Thus, the k_{rw} curve in Figure 4.13(b) must pass through the point $(1,1)$. The excluded data point was incompatible with this condition. Note that the fit to the other points on the k_{rw} curve is extremely good (the mean square residual is 3.08×10^{-3}).

Extension of the two-phase data contained in Figures 4.11-4.13 to the three-phase system is not a straightforward process. As was described

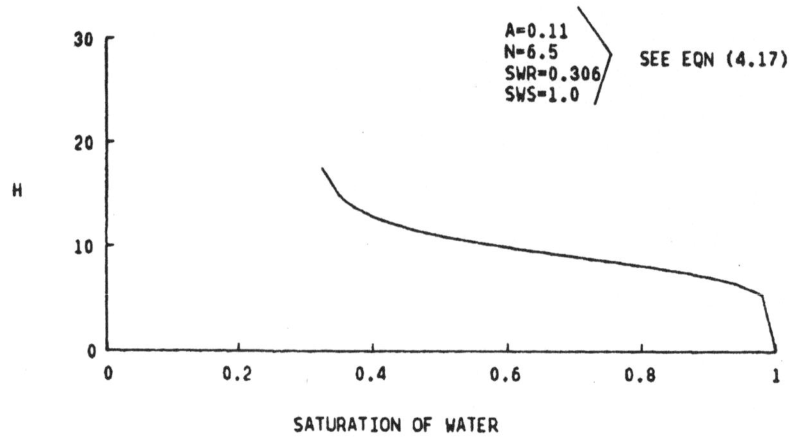

FIGURE 4.13(A): CAPILLARY CURVE FOR WATER-TCE SYSTEM (DRAINAGE)

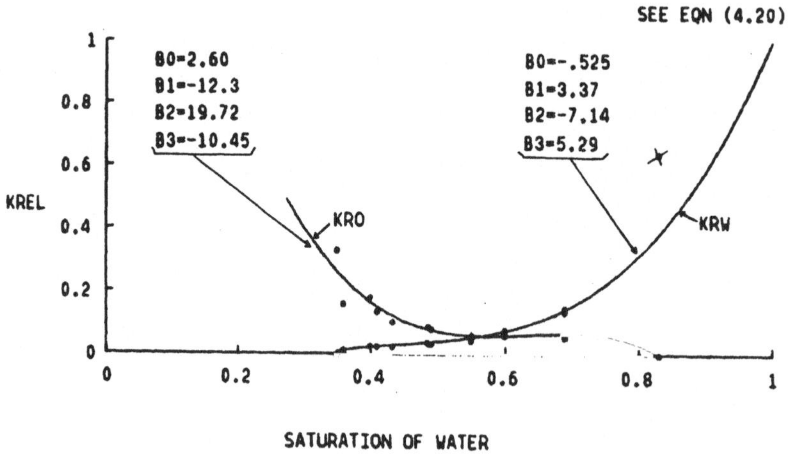

FIGURE 4.13(B): RELATIVE PERMEABILITIES OF WATER AND TCE
(IMBIBITION OF TCE)

in Section 3.4, Stone (1970, 1973) has offered two methods for the prediction of three-phase relative permeabilities from two-phase data. Neither has been adequately field tested. His first method (equations (3.30)-(3.31)) was chosen for this set of TCE simulations. A list of the required parameters and their values is included in Table 4.4.

The evaluation of three-phase saturations from two-phase data has not received much attention in the literature. This is due primarily to the fact that capillary pressures have not shown themselves to be of great physical importance in the description of oil reservoir flows (Shutler (1969)). Most reservoir models have used the following assumptions to handle three-phase saturation relations (Aziz and Settari (1979)):

$$\begin{aligned} s_w &= f(P_{ow}) \\ s_g &= f(P_{og}) \end{aligned} \tag{4.21}$$

Equations (4.21) were suggested by some experimental findings of Leverett and Lewis (1941) but have not been rigorously tested. Note that the water-gas capillary curve is not employed. Thus, in a water dominated system, it does not seem reasonable to use equations (4.21).

In the absence of any accepted method of evaluating three-phase saturations for a groundwater system, the following assumptions are used in this numerical model:

$$\begin{aligned} s_w &= f(P_{ow}) \\ s_g &= f(P_{wg}) \end{aligned} \tag{4.22}$$

Water saturation is, thus, calculated directly from the curve shown in Figure 4.13(a) and gas saturation is determined directly from the curve in Figure 4.12(a) ($s_g = 1 - s_w$). Constraint (2.58) is then employed to calculate the saturation of the organic phase. This scheme is very attractive because of its simplicity. Note also that, in the absence of the gas or organic phases, this method reduces to the appropriate two-phase approach. Various alternative schemes for determining saturations using some weighting of all three capillary relations (Figures 4.11(a) - 4.13(a)) were considered. Such a weighting approach however, was ultimately discarded due to its greater complexity and the lack of three-phase experimental data to support it.

Recall that the saturation relations used for the oil simulations (equations (4.1) and (4.2)) are linear in the pressure variables. For these simulations, then, derivatives of saturations with respect to pressures remain constant. In the TCE case, however, a given saturation expression takes the form (4.17) and calculated derivatives will vary over the domain. Theoretically, fluid saturation derivatives become zero when residual levels of the fluid are reached. That is, in these residual regions, saturations will not change with pressure. Use of zero-valued saturation derivatives, however, may introduce numerical difficulties as described below.

Consider equations (3.2) - (3.4) for the case in which both the matrix and the fluids are incompressible and there is no partitioning of mass. Equation (3.2) becomes:

$$\varepsilon\left[\frac{\partial s_w}{\partial P_{ow}} \frac{\partial P_{ow}}{\partial t} + \frac{\partial s_w}{\partial P_{wg}} \frac{\partial P_{wg}}{\partial t}\right] - \frac{\partial}{\partial x}\left(\tau_w \frac{\partial}{\partial x} P_{wg}\right) + \gamma_w \frac{\partial \tau_w}{\partial x} = 0 \qquad (4.23)$$

Summing equations (3.3) and (3.4) yields the organic phase equation:

$$\varepsilon \left[\frac{\partial s_o}{\partial P_{ow}} \frac{\partial P_{ow}}{\partial t} + \frac{\partial s_o}{\partial P_{wg}} \frac{\partial P_{wg}}{\partial t}\right] - \frac{\partial}{\partial x} \left(\tau_o \frac{\partial}{\partial x} P_{og}\right) + \gamma_o \frac{\partial \tau_o}{\partial x} = 0 \tag{4.24}$$

At nodes for which the organic saturation is below the residual level, $\tau_o = 0$ and equation (4.24) becomes:

$$\frac{\partial s_o}{\partial P_{ow}} \frac{\partial P_{ow}}{\partial t} + \frac{\partial s_o}{\partial P_{wg}} \frac{\partial P_{wg}}{\partial t} = 0 \tag{4.25}$$

If saturation derivatives are permitted to approach zero, equation (4.25) becomes singular.

To safeguard against this eventuality, a small, minimum, value for the saturation derivatives has been incorporated into the numerical simulator. A similar approach has been used successfully by oil reservoir modelers for the simultaneous solution of multiphase equations (Aziz and Settari (1979)). Numerical experiments were conducted with the model to examine the effect of varying the magnitude of this minimum derivative. Values in the range 10^{-5} to 10^{-7} $cm^2/dyne$ were examined. At the low end of this range, convergence of the Newton-Raphson scheme proved difficult. Values approaching 10^{-5} produced significant changes in the solution (in the third decimal place of saturation). A minimum value of about 10^{-6} appears optimum and is used in all the simulations in Section 4.2.

Model diffusion/dispersion parameters for TCE vapor and solute are listed in Table 4.4. In the absence of laboratory data, the value of the dispersivity of TCE in water is merely an order of magnitude estimate from the literature (Freeze and Cherry (1979)). The coefficient for the molecular diffusion of TCE in water was estimated by the method of Hayduk and Laudie (1974). This is the technique recommended by a number of texts (such as Lyman, et al (1982) and Reid, et al (1977)). The Hayduk and Laudie method consists of a correlation expression relating the viscosity of water (μ_W) in cp and molal volume of an organic (V_M') in cm^3/mole to the diffusion coefficient of this organic in water:

$$D^{mw} \simeq \frac{13.26 \times 10^{-5}}{\mu_W^{1.14}(V_M')^{0.589}} \qquad (4.26)$$

Using this equation and the values $\mu_W = 1.0019$ cp and $V_M' = 107.1$ cm^3/mole, D^{mw} becomes 8.434×10^{-6} cm^2/sec.

The molecular diffusion of TCE in air is given in the literature as 0.073 cm^2/sec. Notice that this value is four orders of magnitude larger than the diffusion of TCE in water. In a porous medium, this rate of diffusion will be reduced to some degree by the presence of the soil matrix. Using a capillary tube model, Saffman (1960) estimates that the diffusion of a gas through soil will be about one-third of its value in air. He also cites experimental work with sands that sets the value of this diffusion coefficient at close to two-thirds of its value in air.

Bear and Bachmat (1967) suggest multiplication of the diffusion coefficient by a tortuosity factor. In real applications, however, this factor is difficult to estimate and has been found to vary from 0.01 - 0.5 (Lyman, et al (1982)).

If the diffusion coefficient of one compound in soil is known, the diffusion coefficients of other chemicals can be estimated by the following relation:

$$D_1/D_2 = \sqrt{M_2/M_1} \tag{4.27}$$

where M_2 and M_1 are molecular weights of the compounds and D_1 and D_2 are the corresponding diffusion coefficients. Lyman, et al (1982) use equation (4.27) to estimate the average value of TCE diffusion in soil from known diffusion rates of other organics. They predict a diffusion coefficient of 0.039 cm^2/sec. This value falls within the range which would be anticipated from Saffman's analysis and has been used for the TCE simulations described in this thesis.

Partition coefficients for all TCE simulations were assumed constant in time. Consider the partitioning of TCE into water. At equilibrium, when water is in direct contact with TCE, the solubility of TCE is 1,100 ppm. This is equivalent to 1.509×10^{-4} moles TCE/moles solution. A molar partition coefficient may then be defined as:

$$K_P^W = \frac{\text{moles TCE/moles solution}}{\text{moles TCE/moles organic}} = \frac{1.509 \times 10^{-4}}{x_2^0} \tag{4.28}$$

As long as the denominator in the above equation remains constant (i.e., the composition of the organic phase does not change significantly with time), K_p^w will also be constant. Equation (4.28) was used to evaluate K_p^w for an inflowing organic phase of fixed composition. Table 4.4 lists the value of K_p^w for $x_2^o = 0.5$.

Henry's Law describes the partitioning of mass between the water and gas phases. For TCE at 20°C, Henry's constant is given as (Lyman, et al (1982)):

$$H = \frac{P_{vp}}{s} = 1 \times 10^4 \ \frac{\text{atm} \cdot \text{cm}^3}{\text{mol}} \tag{4.29}$$

Here, s is the solubility of TCE in water and P_{vp} is the partial pressure of TCE in the gas phase at this solubility limit. The partial pressure, which is equivalent to the mole fraction of TCE in the gas phase, may be calculated from (4.29) to be $P_{vp} = x_2^g = 0.08371$. A gas molar partition coefficient may then be computed from its definition as:

$$K_p^g = \frac{x_2^g}{x_2^w} = \frac{0.08371}{1.509 \times 10^{-4}} = 5.549 \times 10^2 \tag{4.30}$$

A sketch of a TCE column experiment is shown in Figure 4.14. The dimensions in the figure are taken from an actual column experiment conducted by Lin (1982). Here, TCE infiltrates under a constant pressure head into a column of water-saturated sand. Water pressure at the lower boundary is maintained at a constant value. Initially, the water

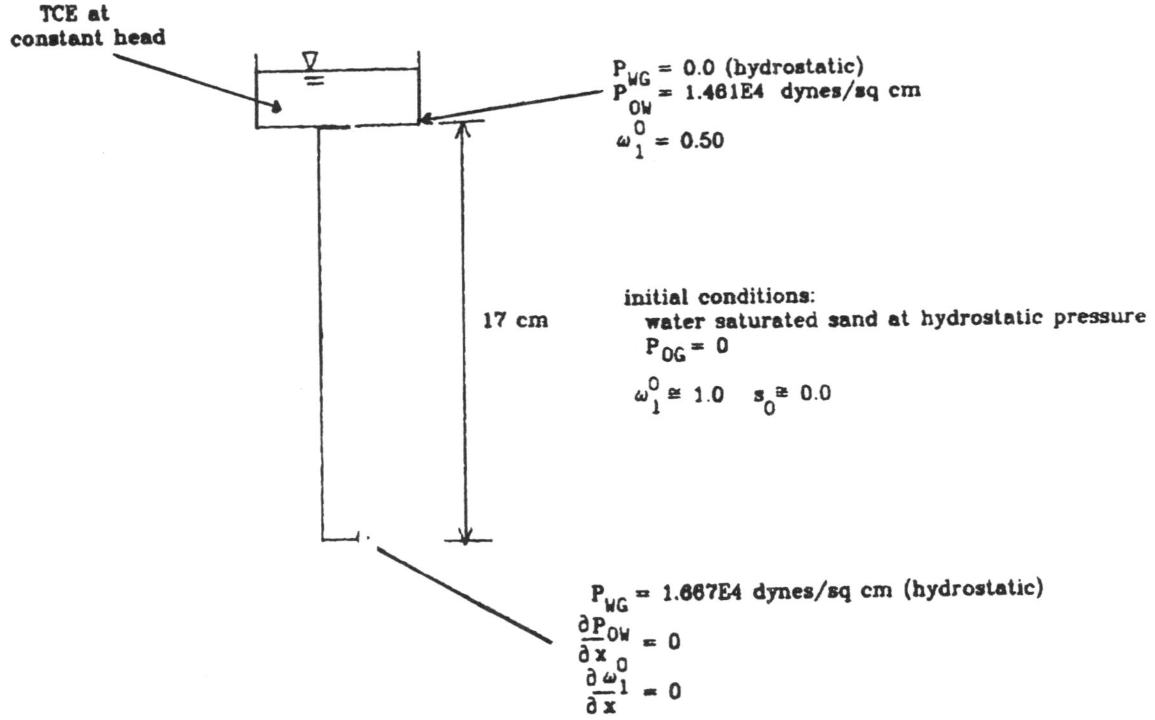

FIGURE 4.14: TCE SIMULATION SCHEMATIC

in the column is in hydrostatic equilibrium. To simulate this experiment, a negligible saturation of TCE is placed at each node at the start of the simulation and P_{og} is set to zero everywhere in the column (see discussion of initial conditions in Section 3.3). The domain is divided into 1 cm increments for a total of 18 nodes. All simulation parameters are included in Table 4.4.

Specification of the boundary condition on P_{ow} at the lower boundary in Figure 4.14 is problematic. For comparison with experimental data, it is desirable to continue the simulation until the organic phase just reaches the boundary. Fortunately, the boundary condition on P_{ow} does not affect the solution as long as the organic phase has not reached the boundary. This can be seen by examining terms involving spatial derivatives of P_{ow} in the mass balance equations (3.3) and (3.4). As long as τ_o is negligible, these terms are also negligible and the solution is unaffected by the choice of capillary pressure gradient. A zero change in capillary pressure across the lower boundary was specified for this simulation.

Notice, then, that at the lower boundary of the column there is a Dirichlet condition on P_{wg} and two Neumann conditions on the other unknowns. Handling boundary conditions of different types at a node presents some difficulty. Recall that a Neumann condition is incorporated into the matrix by writing a full mass balance equation at the boundary node. Because equations for the mass balance of the organic species at the lower boundary contain gradients in P_{wg}, some approximation to these gradients must be sought. For these TCE infiltration simulations, it was assumed that the pressure function P_{wg} is linear at the lower boundary and that the second-derivative of P_{wg} is, thus, zero at that point. Again, as with the spatial derivatives of P_{ow}, it must be emphasized that terms involving the spatial derivatives of P_{wg} do not contribute significantly to the organic species mass balance equations as long as τ_0 and ω_2^w are negligible in the vicinity of the node (that is, as long as the organic phase has not reached the boundary). Thus, the approximation of these derivatives is not critical to the solution until the organic phase has propagated through the column.

Simulation results are presented in Figure 4.15 in the form of saturation profiles of TCE infiltration. Here the curve labels refer to the seconds elapsed from the start of the simulation. A very small initial time step was required ($\sim$ 0.001 sec) to achieve convergence. Thereafter, the time step size was increased until a value of 0.25 sec was reached, and this value was maintained throughout the remainder of the simulation. The maximum TCE saturation of approximately 0.65 at the uppermost node is determined solely by the constant pressure condition at that point.

Flattening of each profile at its downstream end is caused by what is termed the "entry pressure effect". This refers to the pressure (saturation) build-up which must occur at a given node before relative permeability becomes non-zero and flow may proceed to the next node. Comparison of Figure 4.15 with the oil profiles shown in Figure 4.2 reveals that the TCE curves exhibit a markedly different shape. The shape of a saturation profile depends to a great extent on the shape of the corresponding relative permeability curves. For the oil propagation example, the relative permeability of oil to gas (equation (4.5)) is a

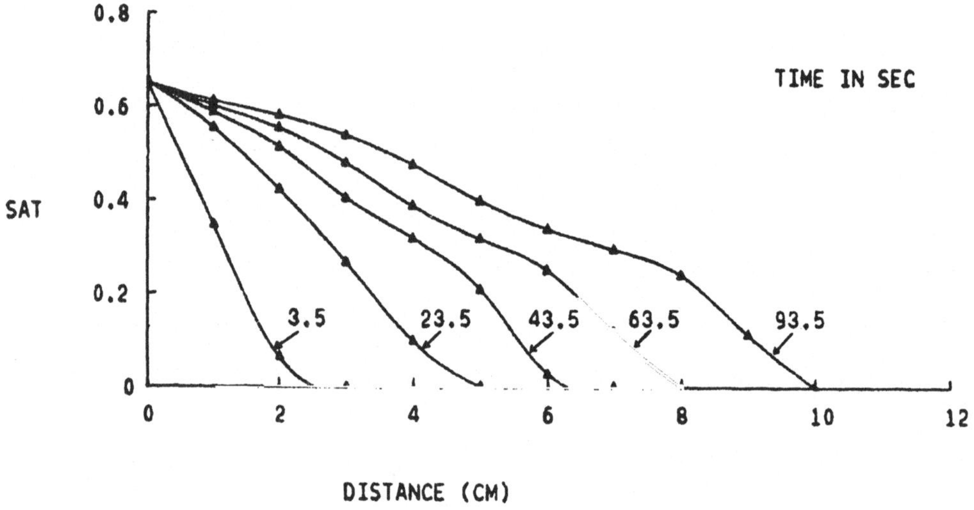

FIGURE 4.15: PROPAGATION OF TCE

quadratic in saturation while the TCE-water relative permeability curve is a cubic. Over a saturation range of about 0.50 to 0.70, TCE permeability changes very little. This behavior is reflected in the apparent "stretching" of the TCE saturation profile in Figure 4.15. In contrast, note that the slope of oil relative permeability varies linearly with saturation throughout the entire saturation range, and the oil saturation profile in Figure 4.2 has a rounder, less elongated shape.

Sample mass conservation results for the TCE simulations are shown in Figure 4.16(a). Here the ratio R_M (see equation (4.13)) has been plotted against time step. Flux was calculated according to equation (4.12) at time level $n+1$ and change in mass by equations (4.10) and (4.11). Recall that a value of $R_M = 1$ is optimal. Results are plotted for the organic phase only but these are virtually identical to those computed for the water phase. This is not surprising, since the flux of TCE into the system should be matched by the flux of water out of the system (neglecting any compressibility effects).

The center plot in Figure 4.16(a) represents typical mass balance results from the simulations used to generate Figure 4.15. Although for any given time step the ratio R_M may dip as low as 0.857 in the figure, the cumulative mass balance ratio is about 0.976. Note that the mass balance plots exhibit a pattern of some regularity. Low values of R_M are found to occur at points in time when the saturation of organic at the leading node approaches its residual value and the relative permeability becomes nonzero. Time step size has little effect on the character of these mass balance plots as long as the step is small enough to permit convergence of the scheme. The variation of high value to low value of

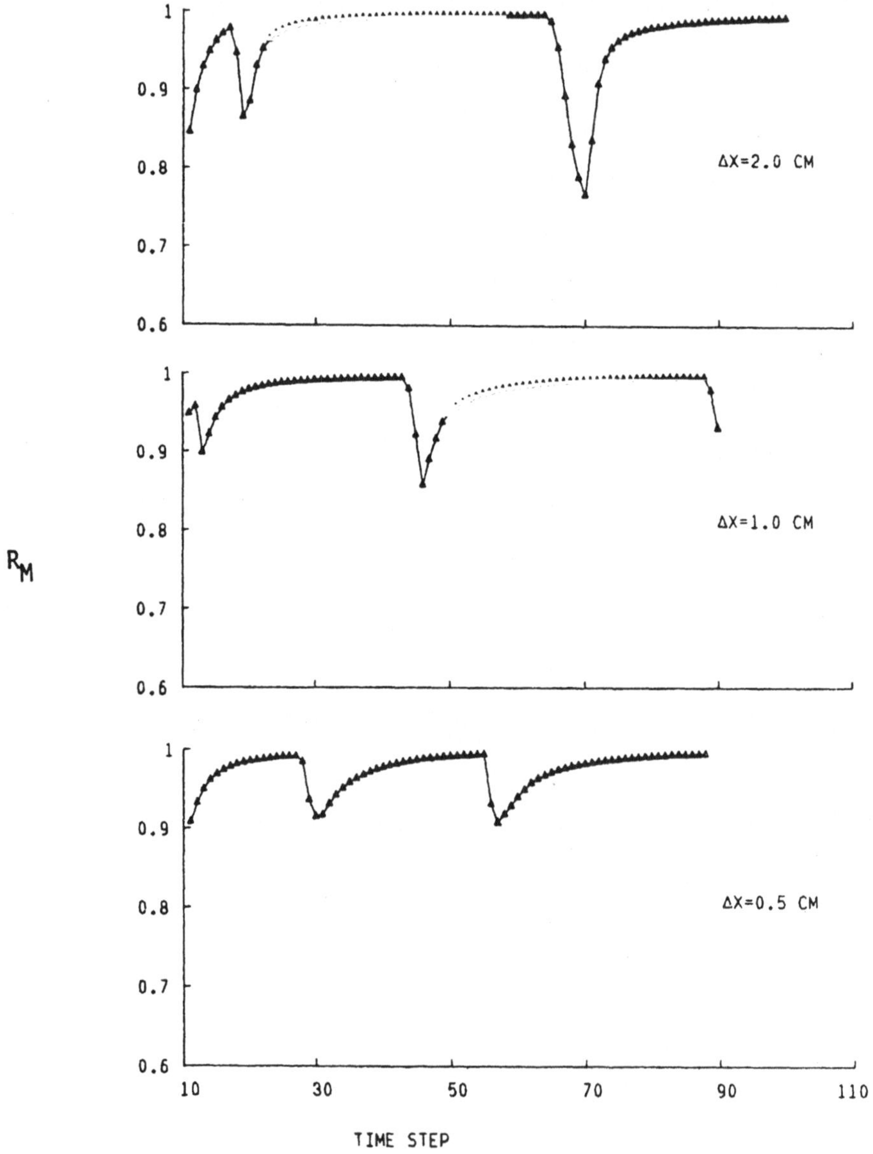

FIGURE 4.16(A): MASS BALANCE FOR DIFFERENT NODAL SPACINGS

R_M, however, is significantly affected by nodal spacing. Consider the other two plots shown in Figure 4.16(a). Notice that for larger nodal spacing, R_M dips as low as 0.767 while the lowest value for the bottom plot (small Δx) is 0.907. Cumulative mass balances are 0.971 and 0.976, respectively, for these two nodal spacings.

The low points on the plots in Figure 4.16(a) also coincide with points of slowest convergence (highest number of iterations) of the Newton-Raphson scheme. Figure 4.16(b) shows the convergence results for two different time steps of a simulation. Here R^{ν} is computed from equation (4.8). Although the solution is converging at both times, each curve exhibits markedly different convergence behavior. The curve labeled (2) in the figure is representative of one of the low points on the mass balance plot. Its convergence rate is a good deal slower than that of curve (1) which represents a point on the mass balance curve approaching $R_M = 1$.

Close examination of Figure 4.13(b) reveals that the slope of the relative permeability curve, k_{ro}, changes abruptly at the value $s_w = 0.83$. It was thought that this abrupt change in slope was causing the convergence behavior exhibited by curve (2) in Figure 4.16(b). Unfortunately, smoothing of the k_{ro} curve by the use of cubic Hermite polynomials failed to improve the convergence properties at this point. Convergence at such "problem points", however, did improve if the time step size was decreased.

Time step size, then, was found to be limited by the iteration requirements at these points of slow convergence. At more "well-behaved" points, the time step size could be increased by a factor of

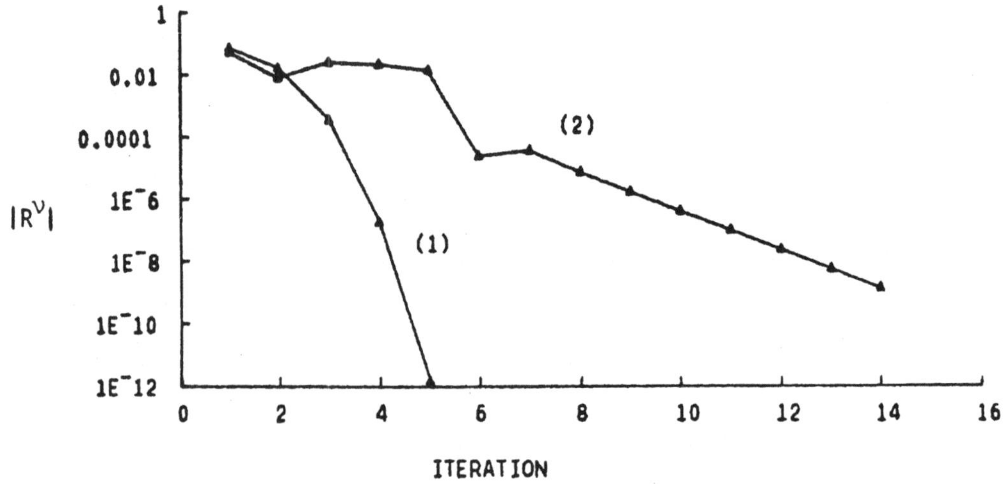

FIGURE 4.16(B): CONVERGENCE OF ITERATION SCHEME

4 or more without encountering convergence problems. Time step size was also greatly affected by the travel time of the organic phase between nodes. If Δx was increased by a factor, the required Δt could also be increased by the same factor. Initial time step size had to be kept very small because of the sudden change in properties at node 1 at time 0^+. It was found that if no partitioning of mass into other phases was permitted, this initial time step could be increased by an order of magnitude.

Propagation of the contaminant plume in the water phase is illustrated in Figures 4.17(a), (b), (c). Saturation profiles of the organic phase at 10, 20, and 30 seconds are given in Figure 4.17(a). Figure 4.17(b) presents the corresponding concentration profiles of TCE in the water phase. Recall that the phase equilibrium assumption requires that the concentration of TCE be at the solubility limit (1100 ppm) when the organic phase is present. This explains the flat portion of the concentration profiles in the figure. TCE is detectable in the water phase out to a distance of about 5 cm beyond the saturation front, due to dispersive/diffusive flux. Concentrations in this zone are better illustrated in the semi-log plot of Figure 4.17(c). Note that the concentrations span about four orders of magnitude over the length of the column. The value of 2.2×10^{-8} (22 ppb) on the ordinate axis in the figure is the background level of TCE present at all the nodes at the start of the simulation. (This concentration level corresponds to $\omega_2^o = 0.00001$).

The effect of organic phase upstream weighting on the solution is shown in Figure 4.18. Here the use of upstream weighting tends to

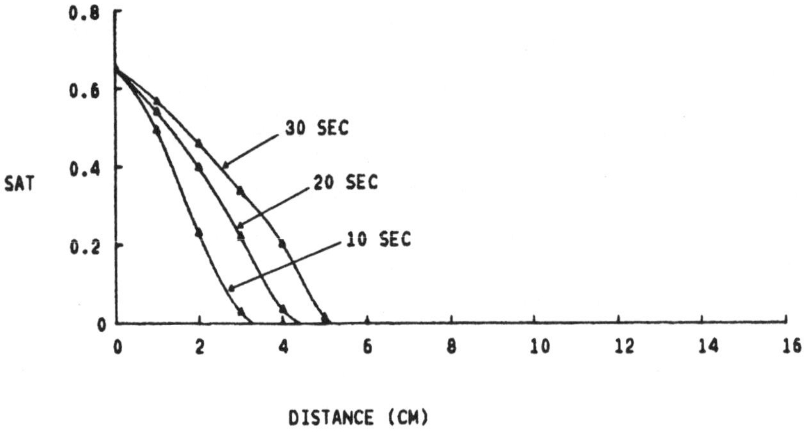

FIGURE 4.17(A): INFILTRATION OF TCE

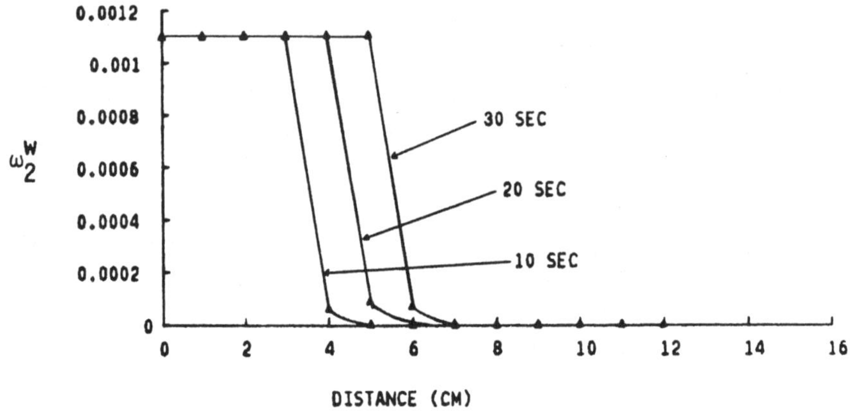

FIGURE 4.17(B): CONCENTRATION IN THE WATER PHASE

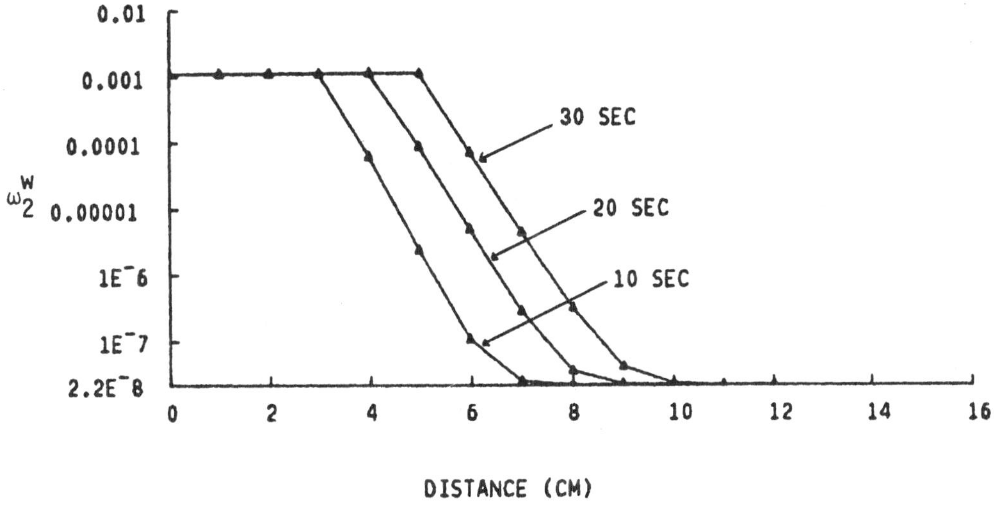

FIGURE 4.17(C): SEMI-LOG CONCENTRATION PLOT

smear the front slightly but does not shift its location as in the oil problem (see Figure 4.7(a)). In the upstream weighted case, mass balance results on the organic phase tended to be poorer (R_M as low as 0.80) than those on the water phase. Thus, it appears that upstream weighting offers no improvement to the solution of this problem.

A laboratory experiment similar to that shown in Figure 4.14 was conducted by Lin (1982). In this experiment, however, the lower end of the column was open to the atmosphere ($P_{wg} = 0.0$ at the lower end). Lin recorded the time it took for the TCE to travel the length of the column and to appear at the lower end. Over the course of the experiment, the head of TCE at the upper end of the column varied by as much as 10% of its average value in either direction. Unfortunately, this variation was not recorded. The average head value maintained is that shown in Figure 4.14. A TCE travel time of 102 seconds was measured.

Using a boundary condition of $P_{wg} = 0.0$ at the lower end of the column, a simulation was carried out by the computer model to determine the approximate travel time of the TCE. Partitioning of TCE into the other phases was neglected in these runs. Figure 4.19 shows the effect of the changed boundary condition on the propagation of the front. The two saturation profiles shown in the figure correspond to simulations using two different boundary conditions at the lower end. Both of these profiles represent a simulated time of 8.5 seconds. Note that TCE propagates much fastem in the column open to the atmosphere and that the saturation profile of this simulation is also more rectangular in shape.

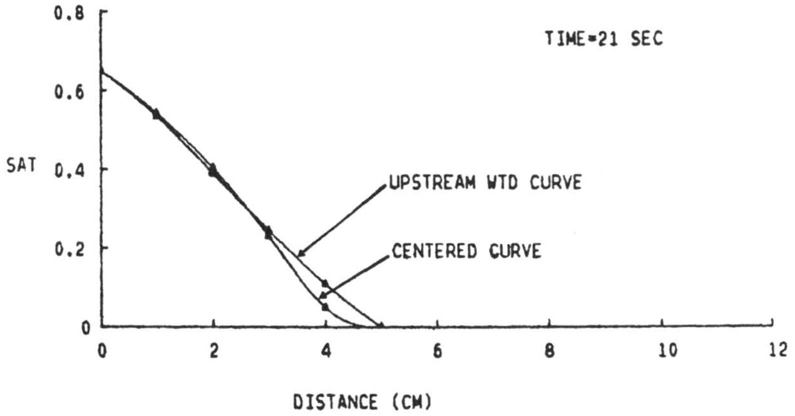

FIGURE 4.18: UPSTREAM WEIGHTING COMPARISON

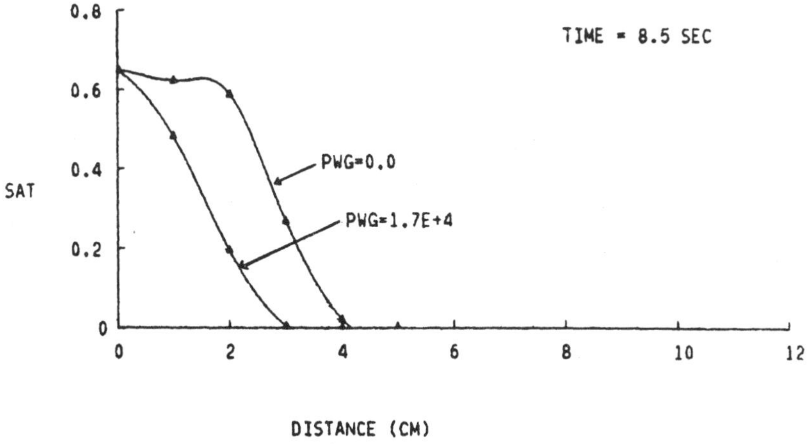

FIGURE 4.19: EFFECT OF BOUNDARY CONDITION AT BOTTOM NODE

In the travel time simulation, TCE first reached the boundary node after 100.5 seconds of simulation time. At a time of 106.0 seconds, the concentration of TCE reached the residual level at this node. It is not clear which of these two times corresponds to the breakthrough time measured by Lin. Considering all the uncertainties regarding boundary conditions, these two times are in surprisingly good agreement with the experimental measurement of 102 seconds.

CHAPTER V

THE TWO-DIMENSIONAL SIMULATOR

This chapter deals with the extension of the one-dimensional model described in Chapter III to two space dimensions. Section 5.1 discusses the development of the appropriate difference equations and the incorporation of boundary conditions. Matrix structure and model solution procedure are examined in Section 5.2. The final section of this chapter presents an example two-dimensional simulation of TCE migration in a confined aquifer.

5.1 Extension to Two Space Dimensions

The mass balance equations governing the multiphase flow of fluids for the two-dimensional (2-D) model are given by equations (2.55) - (2.57) where the gradient operator is defined as:

$$\underset{\sim}{\nabla}(\;) \equiv \frac{\partial}{\partial x}(\;)\underset{\sim}{i} + \frac{\partial}{\partial y}(\;)\underset{\sim}{j} \qquad (5.1)$$

Here $\underset{\sim}{i}$ and $\underset{\sim}{j}$ are two unit vectors in the coordinate directions x and y respectively. The gravity terms appearing in the mass balance equations may be expanded as follows:

$$\rho^{\alpha} g \nabla z \equiv \rho^{\alpha} g \cos \lambda_x \, i + \rho^{\alpha} g \cos \lambda_y \, j \qquad (5.2)$$

where λ_x and λ_y are the angles which the x and y coordinate directions make with the vertical (measured in a clockwise direction). As with the 1-D model, adsorption is neglected in equation (2.57) due to lack of experimental data (see Section 3.2).

For simplification, it is assumed in the model that the principle axes of the intrinsic permeability tensor, k, coincide with the coordinate directions. This is a reasonable assumption for groundwater flow problems if the coordinate axes are taken perpendicular and parallel to the bedding planes. Under these conditions, the permeability tensor is given by:

$$k = \begin{bmatrix} k_x & 0 \\ 0 & k_y \end{bmatrix} \qquad (5.3)$$

Terms of the form $\nabla \cdot (A\tau \cdot \nabla P)$ which appear in equations (2.55)-(2.57) may then be expanded as:

$$\nabla \cdot (A\tau \cdot \nabla P) = \frac{\partial}{\partial x} \left(A \, \tau_x \frac{\partial P}{\partial x}\right) + \frac{\partial}{\partial y} \left(A \, \tau_y \frac{\partial P}{\partial y}\right) \qquad (5.4)$$

Here, τ_x and τ_y are the non-zero components of the tensor τ. Note that no cross-derivative terms appear in the expansion (5.4).

Formation of the difference equations proceeds in a manner analogous to that described in Section 3.2. Time derivative expansions are identical to those given by (3.9), (3.10), (3.11), and (3.14). Each derivative expansion in x is matched by a corresponding expansion in y. Thus, spatial derivative approximations are given as:

$$\left\{\begin{aligned} \frac{\partial}{\partial x} f\Big|_{i,j} &\simeq \frac{1}{\Delta x_+ + \Delta x_-}\left[f_{i,j+1} - f_{i,j-1}\right] \\ \frac{\partial}{\partial y} f\Big|_{i,j} &\simeq \frac{1}{\Delta y_+ + \Delta y_-}\left[f_{i+1,j} - f_{i-1,j}\right] \end{aligned}\right. \tag{5.5}$$

$$\left\{\begin{aligned} \frac{\partial}{\partial x}\left(\Omega \frac{\partial f}{\partial x}\right)\Big|_{i,j} &\simeq \frac{2}{\Delta x_+ + \Delta x_-}\Big[\Omega_{i,j+\frac{1}{2}} \frac{f_{i,j+1} - f_{i,j}}{\Delta x_+} \\ &\qquad - \Omega_{i,j-\frac{1}{2}} \frac{f_{i,j} - f_{i,j-1}}{\Delta x_-}\Big] \\ \frac{\partial}{\partial y}\left(\Omega \frac{\partial f}{\partial y}\right)\Big|_{i,j} &\simeq \frac{2}{\Delta y_+ + \Delta y_-}\Big[\Omega_{i+\frac{1}{2},j} \frac{f_{i+1,j} - f_{i,j}}{\Delta y_+} \\ &\qquad - \Omega_{i-\frac{1}{2},j} \frac{f_{i,j} - f_{i-1,j}}{\Delta y_-}\Big] \end{aligned}\right. \tag{5.6}$$

where

$$\Omega_{i,j\pm\frac{1}{2}} \equiv \frac{1}{2}(\Omega_{i,j} + \Omega_{i,j\pm 1})$$

and

$$\Omega_{i\pm\frac{1}{2},j} \equiv \frac{1}{2}(\Omega_{i,j} + \Omega_{i\pm 1,j})$$

Here, the subscripts i,j indicate the row and column location of a node. Note that the above difference expressions are analogous to (3.5)-(3.6) and have the same properties (see Section 3.2). A sample finite difference grid of dimension 4×6 is shown in Figure 5.1. Coordinate directions and nodal spacings used in the expansions (5.5) and (5.6) are indicated in the figure.

As with the 1-D model, an upstream weighting option has been incorporated into the 2-D simulator. Mobilities $\tau_{\alpha x}$ and $\tau_{\alpha y}$ may be weighted in space in the x or y coordinate directions respectively:

$$\begin{aligned}
(\tau_{\alpha x})_{i,j+\frac{1}{2}} &= \theta_{\alpha x}(\tau_{\alpha x})_{i,j} + (1 - \theta_{\alpha x})(\tau_{\alpha x})_{i,j+1} \\
(\tau_{\alpha x})_{i,j-\frac{1}{2}} &= \theta_{\alpha x}(\tau_{\alpha x})_{i,j-1} + (1 - \theta_{\alpha x})(\tau_{\alpha x})_{i,j} \\
(\tau_{\alpha y})_{i+\frac{1}{2},j} &= \theta_{\alpha y}(\tau_{\alpha y})_{i,j} + (1 - \theta_{\alpha y})(\tau_{\alpha y})_{i+1,j} \\
(\tau_{\alpha y})_{i-\frac{1}{2},j} &= \theta_{\alpha y}(\tau_{\alpha y})_{i-1,j} + (1 - \theta_{\alpha y})(\tau_{\alpha y})_{i,j}
\end{aligned} \tag{5.7}$$

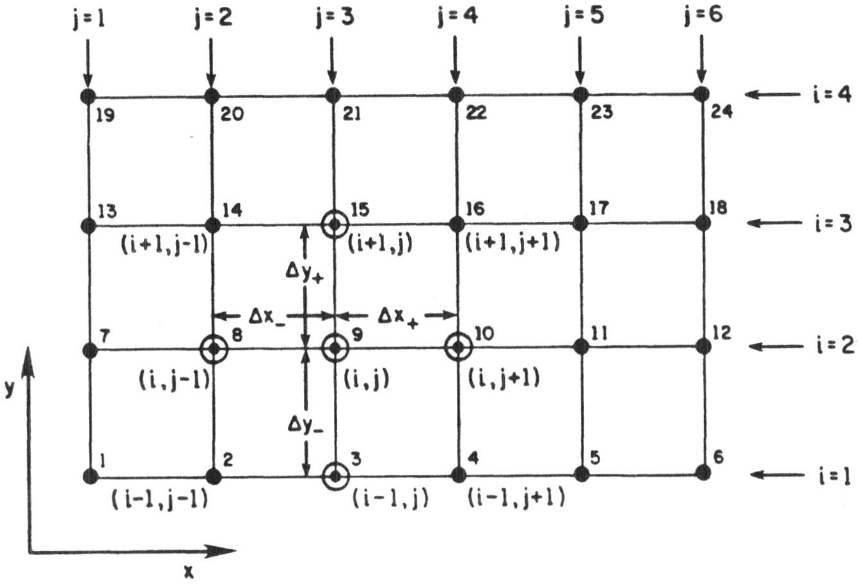

FIGURE 5.1: SAMPLE 2-D GRID

Here $0 \leq \theta_{\alpha x} \leq 1$ and $0 \leq \theta_{\alpha y} \leq 1$ are upstream weighting parameters input to the program. Note that equations (5.7) are analogous to equations (3.8).

In two space dimensions, the dispersivity tensor, a^{α}_{ijkm}, (described in Section 3.4) is related to two porous medium constants, the longitudinal dispersivity, a^{α}_{L}, and the transversal dispersivity, a^{α}_{T} (Bear (1979)). Using these two constants, the dispersivity tensor may be written as:

$$a^{\alpha}_{ijkm} = a^{\alpha}_{T}\delta_{ij}\delta_{km} + \frac{a^{\alpha}_{L} - a^{\alpha}_{T}}{2}(\delta_{ik}\delta_{jm} + \delta_{im}\delta_{jk}) \tag{5.8}$$

where δ_{ik} is the Kronecker delta. Substitution of (5.8) into the dispersion tensor expression (3.19a) yields the following form for the dispersion tensor in two space dimensions:

$$D^{\alpha}_{ij} = \begin{bmatrix} D^{m\alpha} + (a^{\alpha}_{T}\bar{v}^{\alpha^2}_{y} + a^{\alpha}_{L}\bar{v}^{\alpha^2}_{x})/\bar{v}^{\alpha} & (a^{\alpha}_{L} - a^{\alpha}_{T})\bar{v}^{\alpha}_{x}\bar{v}^{\alpha}_{y}/\bar{v}^{\alpha} \\ (a^{\alpha}_{L} - a^{\alpha}_{T})\bar{v}^{\alpha}_{x}\bar{v}^{\alpha}_{y}/\bar{v}^{\alpha} & D^{m\alpha} + (a^{\alpha}_{T}v^{\alpha^2}_{x} + a^{\alpha}_{L}\bar{v}^{\alpha^2}_{y})/\bar{v}^{\alpha} \end{bmatrix} \tag{5.9}$$

Here, $D^{m\alpha}$ is the molecular diffusion coefficient for the medium, $\bar{v}^{\alpha}$ is the average fluid velocity, and $\bar{v}^{\alpha}_{x}$ and $\bar{v}^{\alpha}_{y}$ are the magnitudes of the velocity components in the coordinate directions. Relation (5.9) is the form of the dispersion tensor used in the simulator for dispersion in the organic and water phases. The constant parameters $D^{m\alpha}$, a^{α}_{T}, and

a_L^{α} must be input to the model. Velocity components $\bar{v}_x^{\alpha}$ and $\bar{v}_y^{\alpha}$ are calculated internally by a finite difference discretization of Darcy's law (equation (2.25)). Dispersion coefficients are updated at the end of each time step.

Expansion and discretization of the dispersion terms in equations (2.55) - (2.57) require special consideration. Terms of the form $\underset{\sim}{\nabla}\cdot(A\underset{\approx}{D}\cdot\underset{\sim}{\nabla}\omega)$ may be expanded in two space dimensions as:

$$\underset{\sim}{\nabla}\cdot(A\underset{\approx}{D}\cdot\underset{\sim}{\nabla}\omega) = \frac{\partial}{\partial x}(A\,D_{xx}\frac{\partial\omega}{\partial x} + A\,D_{xy}\frac{\partial\omega}{\partial y}) + \frac{\partial}{\partial y}(A\,D_{yx}\frac{\partial\omega}{\partial x} + A\,D_{yy}\frac{\partial\omega}{\partial y}) \tag{5.10}$$

where D_{xx}, D_{xy}, D_{yx}, and D_{yy} are given by (5.9). Cross derivative terms will appear in (5.10) as long as the direction of fluid velocity does not coincide with a coordinate direction. These cross derivative terms are approximated within the model by the following finite difference expressions:

$$\left.\frac{\partial}{\partial y}(\Omega\frac{\partial f}{\partial x})\right|_{i,j} \simeq \frac{1}{\Delta y_+ + \Delta y_-}\left[\Omega_{i+1,j}\frac{f_{i+1,j+1} - f_{i+1,j-1}}{\Delta x_+ + \Delta x_-} - \Omega_{i-1,j}\frac{f_{i-1,j+1} - f_{i-1,j-1}}{\Delta x_+ + \Delta x_-}\right]$$

$$\left.\frac{\partial}{\partial x}(\Omega\frac{\partial f}{\partial y})\right|_{i,j} \simeq \frac{1}{\Delta x_+ + \Delta x_-}\left[\Omega_{i,j+1}\frac{f_{i+1,j+1} - f_{i-1,j+1}}{\Delta y_+ + \Delta y_-} - \Omega_{i,j-1}\frac{f_{i+1,j-1} - f_{i-1,j-1}}{\Delta y_+ + \Delta y_-}\right] \tag{5.11}$$

Other terms in (5.10) are discretized according to (5.6).

Examination of the finite difference expansions (5.5) and (5.6) reveals that they involve unknowns at a maximum of five spatial locations. (These locations are circled in Figure 5.1). Incorporation of (5.11) into the model adds unknowns at four additional points:

$$(i+1,j+1), \quad (i+1,j-1), \quad (i-1,j-1) \quad (i-1,j+1)$$

To avoid the resultant increase in bandwidth and computation time that the addition of these four points would entail, the cross-derivative terms (5.11) are lagged by one iteration in the model. Thus, they appear in the right-hand side vector only.

Boundary conditions for the two-dimensional model are treated in a manner analogous to the one-dimensional case as presented in Section 3.3. Along a side, the boundary conditions may be of Dirichlet or Neumann type. For Dirichlet conditions, the mass balance equation at a node is replaced by an identity of the form (3.16). In the case of a constant normal gradient in the x or y direction, a value of the variable u at an "imaginary node" is calculated by:

$$u_{i,0} = u_{i,2} - 2\Delta x_1 \left(\frac{\partial u}{\partial x}\right)_{i,1}$$
$$u_{i,NCOL+1} = u_{i,NCOL-1} + 2\Delta x_{NCOL} \left(\frac{\partial u}{\partial x}\right)_{i,NCOL} \qquad (5.12a)$$

or

$$u_{0,j} = u_{2,j} - 2\Delta y_1 \left(\frac{\partial u}{\partial y}\right)_{1,j}$$
$$u_{NROW+1,j} = u_{NROW-1,j} + 2\Delta y_{NROW} \left(\frac{\partial u}{\partial y}\right)_{NROW,j} \qquad (5.12b)$$

Here, subscripts refer to the row and column location. NCOL is the number of columns and NROW is the number of rows in the grid. Discretized mass balance equations are then written at a constant gradient boundary node by incorporation of the appropriate imaginary node values from equations (5.12a) and (5.12b).

5.2 Matrix Equation Structure and Solution

Expansion of the derivatives in equations (2.55) - (2.57) according to the difference expressions presented in Section 5.1 yields a system of nonlinear algebraic difference equations in the unknown pressures P_{ow} and P_{wg} and the unknown mass fractions ω_1^o, ω_2^o, ω_2^w, and ω_2^g. As in the development of the 1-D model, incorporation of the equilibrium relations (2.54) and constraint (2.2) reduces this system to a system of equations in three unknowns: P_{ow}, P_{wg} and ω_1^o. For a summary of the time levels of evaluation and the functional dependence of the coefficients appearing in these equations, consult Table 3.1. Recall that $\underset{\approx}{k}$ and $\underset{\approx}{D}^{\alpha}$ are tensors in the two-dimensional case. Section 3.4 of Chapter III describes the functional forms used in the model for all the scalar parameters.

Let NN be the number of nodes in the discretized domain shown in Figure 5.1. Then the system of nonlinear algebraic difference equations which expresses conservation of mass over this domain may be written in matrix form as:

$$\underset{\approx}{A} \cdot \underset{\sim}{u} = \underset{\sim}{B} \tag{5.13}$$

where $\underset{\approx}{A}$ is a coefficient matrix of size $(3 \times NN) \times (3 \times NN)$, $\underset{\sim}{B}$ is a $3 \times NN$ vector and $\underset{\sim}{u}$ is an ordered vector of $3 \times NN$ unknowns.

Matrix $\underset{\approx}{A}$ is a banded matrix whose structure is shown in Figure 5.2. This structure corresponds to the node numbering scheme used in Figure 5.1. Along the diagonal, there is a block tridiagonal band like that which appears in the coefficient matrix of the 1-D model (see Figure 3.2). The other two off-diagonal bands in the matrix shown in Figure 5.2 are associated with neighboring nodes on adjoining rows in the grid (nodes (i+1,j) and (i-1,j)). Note that each group of three equation spans, at most, five nodes corresponding to the circled nodes in Figure 5.1.

The components of a block subdivision of $\underset{\approx}{A}$, $[A_{kj}]_i$ (k=1,3, j=1,15) are very much like those in the 1-D model (see Appendix C.1). A_{kj} (k=1,3, j=1,9) in the two-dimensional formulation are identical to their 1-D counterparts. A_{kj} (k=1,3, j=10,13) have the same form as A_{kj} (k=1,3, j=1,3) with y-directional parameters replacing x-directional parameters. Similarly, A_{kj} (k=1,3, j=13,15) have the same structure as A_{kj} (k=1,3, j=7,9). All directional (spatial derivative) terms which appear in the 1-D tridiagonal coefficients A_{kj} (k=1,3, j=4,6) and the right-hand side vector B_k (k=1,3) are duplicated in the y-direction in the 2-D version of these coefficients. In addition, B_2 and B_3 contain cross-derivative terms of the form (5.11) relating to dispersion in the two-dimensional system.

The nonlinear matrix equation (5.13) must be solved for the $3 \times NN$ vector of unknowns, $\underset{\sim}{u}$, at each time step. As with the 1-D model, the 2-D simulator employs the Newton-Raphson iteration method to

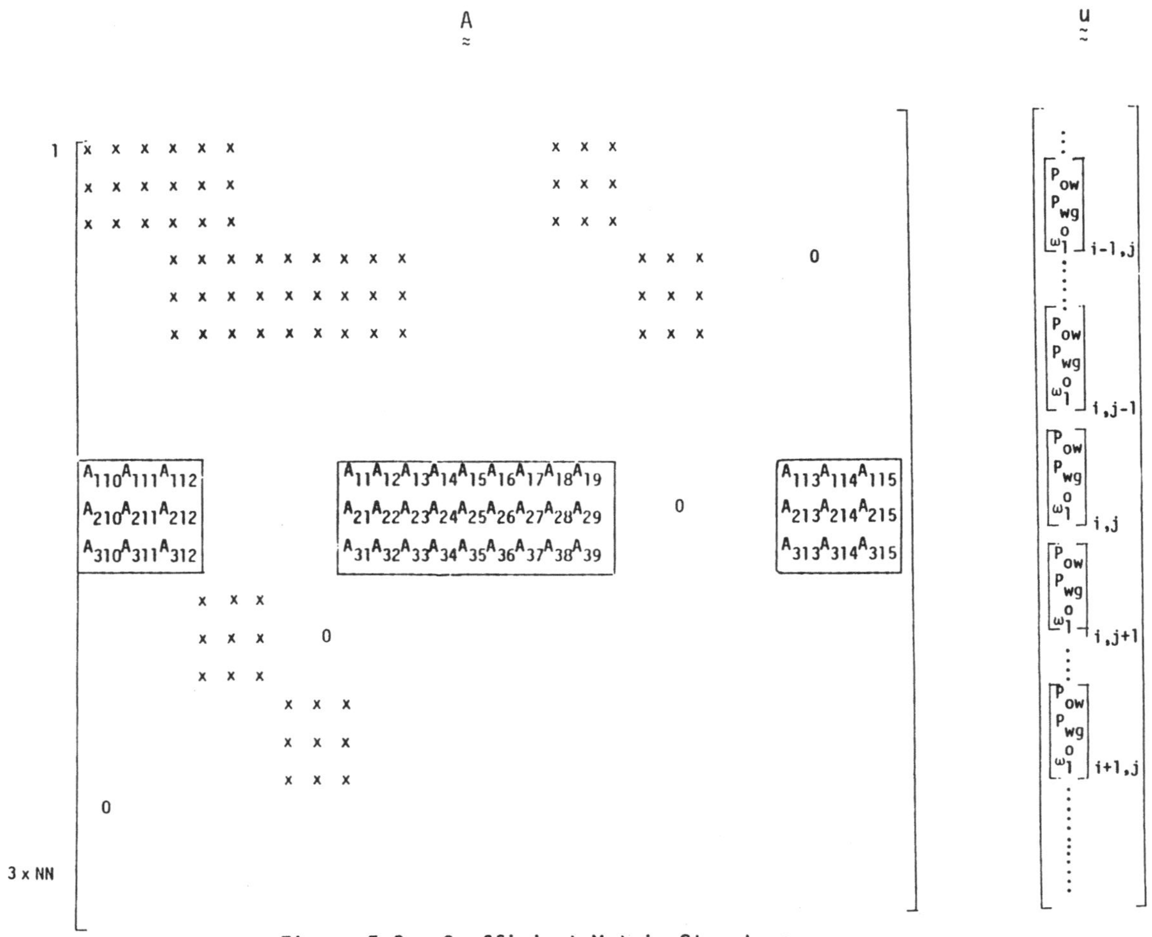

Figure 5.2: Coefficient Matrix Structure

solve this system of equations. For a discussion of this solution technique and the method of derivative evaluation, see Section 3.5. Note that the Newton-Raphson matrix has the same structure as that of the matrix $\underset{\approx}{A}$ shown in Figure 5.2.

Although a band solver could be used directly to solve the Newton-Raphson matrix equations (as in the 1-D model), a more efficient solution method which takes account of the particular matrix structure shown in Figure 5.2 is employed in the 2-D simulator. This method involves renumbering the nodes according to what is known as a "D4" numbering scheme. The oil industry has made use of this renumbering approach when solving systems of equations with a similar matrix structure (see, for example, Aziz and Settari (1979) and Price and Coats (1974)).

Consider the node numbering scheme shown in Figure 5.3(a). Here nodes are numbered consecutively along every other grid diagonal. The matrix structure which corresponds to this numbering system in shown in Figure 5.3(b). In this figure, each block designates a 3×3 matrix. Note that the reordered equations have the following structure:

$$\underset{\approx}{F} \cdot \underset{\sim}{\delta} = \begin{bmatrix} \underset{\approx}{F}^1 & \underset{\approx}{F}^2 \\ \underset{\approx}{F}^3 & \underset{\approx}{F}^4 \end{bmatrix} \cdot \begin{bmatrix} \underset{\sim}{\delta}^1 \\ \underset{\sim}{\delta}^2 \end{bmatrix} = \begin{bmatrix} \underset{\sim}{f}^1 \\ \underset{\sim}{f}^2 \end{bmatrix} \tag{5.14}$$

where $\underset{\approx}{F}^1$ and $\underset{\approx}{F}^4$ are block diagonal matrices and $\underset{\approx}{F}^2$ and $\underset{\approx}{F}^3$ are sparse banded matrices.

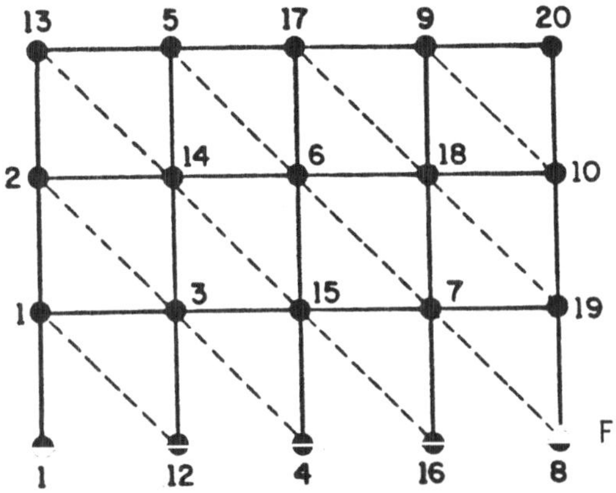

FIGURE 5.3(a): D4 NUMBERING SCHEME

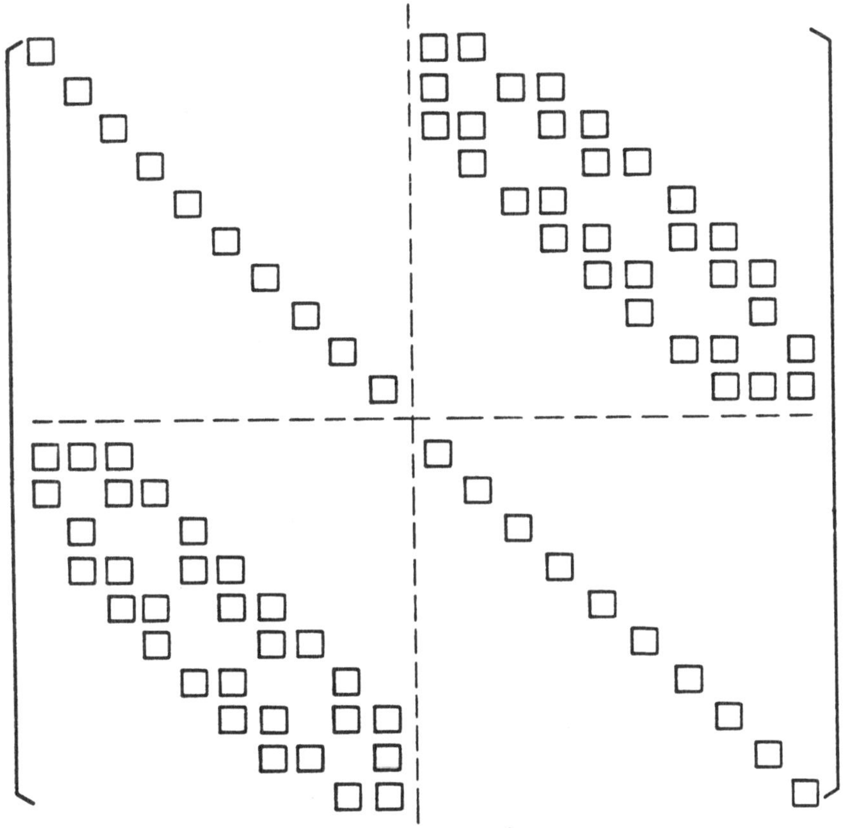

FIGURE 5.3(b): RENUMBERED MATRIX STRUCTURE

The D4 solution procedure may be described briefly as follows. First, the model renumbers nodes according to the D4 scheme shown in Figure 5.3(a). Next, forward elimination is used on the lower half of the $\underset{\approx}{F}$ matrix to transform $\underset{\approx}{F}^3$ into the null matrix. The following matrix structure results:

$$\begin{bmatrix} \underset{\approx}{F}^1 & \underset{\approx}{F}^2 \\ \underset{\approx}{0} & \underset{\approx}{\bar{F}}^4 \end{bmatrix} \cdot \begin{bmatrix} \underset{\sim}{\delta}^1 \\ \underset{\sim}{\delta}^2 \end{bmatrix} = \begin{bmatrix} \underset{\sim}{f}^1 \\ \underset{\sim}{\bar{f}}^2 \end{bmatrix} \tag{5.15}$$

Here a bar designates that a given component has been modified through the forward elimination process. A band solver is next used to compute the solution to the reduced matrix equation:

$$\underset{\approx}{\bar{F}}^4 \cdot \underset{\sim}{\delta}^2 = \underset{\sim}{\bar{f}}^2 \tag{5.16}$$

The upper vector of unknowns, $\underset{\sim}{\delta}^1$ may then be computed directly from:

$$\underset{\sim}{\delta}^1 = {\underset{\approx}{F}^1}^{-1} \cdot \underset{\sim}{f}^1 - {\underset{\approx}{F}^1}^{-1} \cdot \underset{\approx}{F}^2 \cdot \underset{\sim}{\delta}^2 \tag{5.17}$$

Finally, the solution is returned to the old node numbering scheme and unknowns are updated.

Note that the matrix equation (5.16) which is solved directly using this D4 procedure is half the size of that which must be solved directly if a row by row numbering scheme is employed. Further reduction in storage requirements is also made in the D4 solver by storing only non-zero values of the sparse banded matrix $\underset{\approx}{F}^2$. A comparison of the storage and execution time requirements for the 2-D simulator using the two alternative solution schemes is given in Table 5.1 for various grid sizes. Note that, as the grid becomes larger, the D4 scheme offers increased advantage over the direct band solver method.

TABLE 5.1

Solver Comparison

Grid Size	No. of Iterations	Largest Array		Execution time (seconds)		% reduction (time)
		Band	D4	Band	D4	
4 x 4	3	48 x 29	24 x 29	0.83	0.82	1.2
4 x 6	3	72 x 41	36 x 29	1.11	0.99	10.8
4 x 6	5			1.45	1.25	13.8
4 x 10	3	120 x 65	60 x 29	2.10	1.26	40.0
4 x 10	5			3.01	1.75	41.9

With the exception of the solver routine, the 2-D model is structured identically to the 1-D simulator. A flow chart of this structure is given in Figure 3.3. All parameters and unknowns are input and output according to a row by row numbering system. The grid must be numbered in such a way that NCOL $\geq$ NROW.

5.3 Example Simulations

For verification of the 2-D model, the 1-D TCE infiltration problem depicted in Figure 4.14 and discussed in Chapter IV was simulated by the 2-D model. A 40 node grid (4 rows x 10 columns) was superimposed upon a soil column of length 18 cm and width 3 cm. Nodes in the horizontal direction were spaced at 1 cm intervals and in the vertical direction at 2 cm intervals. For boundary conditions along the sides of the column, the horizontal gradient in both the TCE and water pressures was set to zero. Hydrostatic water pressure was maintained at the lower boundary. For a listing of other boundary conditions and simulation parameters, see Figure 4.14 and Table 4.4.

Comparison of the 2-D simulation results with the 1-D predictions showed agreement to the fifth decimal place in the solutions. As anticipated, the 2-D solution profile was uniform across the column. The grid was rotated 90° and the simulation repeated with identical results. (In the first simulation, gravity acted in the y-coordinate direction, and in the second, gravity was oriented in the x-direction.)

Consider next the full two-dimensional problem scenario depicted in Figure 5.4. Here, TCE infiltrates into a sloping confined aquifer through a rift in the upper confining bed. The aquifer is 2.5 m thick, and the region of simulation extends 26 m laterally. The beds are inclined at an angle of 0.01 radians to the horizontal. Model discretization of the region is shown in the figure. The domain consists of 66 nodes (6 rows x 11 columns). Nodes are equally spaced in the y-direction. Spacing in the x-direction increases towards the boundaries to reduce the effect of the lateral boundary conditions on the solution.

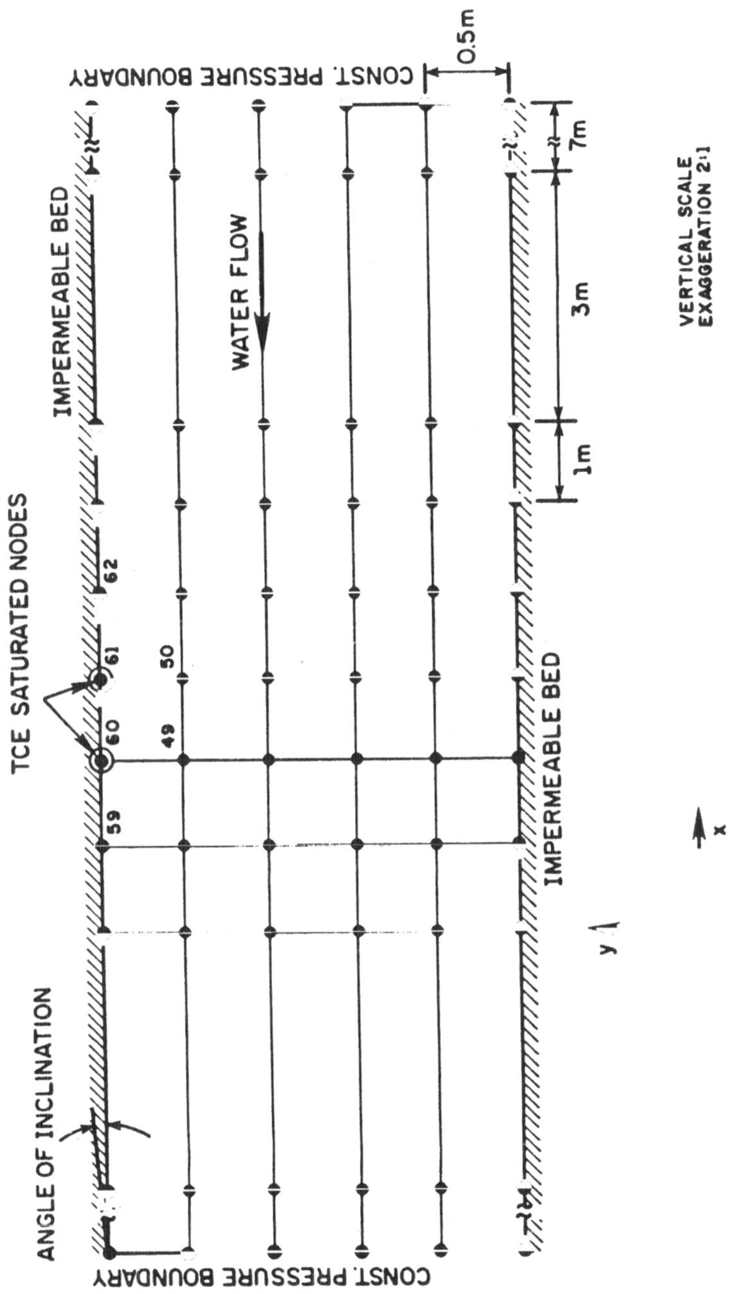

FIGURE 5.4: 2-D SIMULATION SCHEMATIC

Boundary and initial conditions for the problem are listed in Table 5.2. Initially, the nodal water pressures (P_{wg_i}) are in equilibrium. In the vertical direction, there is a hydrostatic distribution of pressure. A horizontal pressure head gradient of 0.00385 m/m is superimposed upon this hydrostatic pressure field to produce an up-slope flow of water. At the start of the simulation, the pressure of the organic phase (P_{og}) is set to zero everywhere. The upper and lower boundaries of the aquifer are assumed impermeable. Saturation and pressure of TCE at the two source nodes remains constant throughout the simulation. Note that, at the right and left boundaries, a zero gradient condition is placed on the mass fraction. These boundaries have been placed far enough away from the polluted area so that this condition does not influence the solution during the period of simulation.

Simulation parameters are summarized in Table 5.3. Most values are identical to those used in the 1-D simulations. Note that a transversal dispersivity value of 0.02 cm is assumed. This is one-fifth of the longitudinal value and within the range of values encountered in experimental work (see Freeze and Cherry (1979)).

Prior to the start of the simulation, the model was run with the initial conditions as the boundary conditions to assure that the starting pressures were in equilibrium. Only one iteration was required to achieve convergence for this problem in the first time step. Initial residuals were on the order of 10^{-16} for each time step thereafter, with no appreciable change in pressures.

Some simulation results are plotted in Figures 5.5(a), (b), and (c). Here, linear interpolation between nodal values was used to develop

TABLE 5.2

Boundary and Initial Conditions for 2-D Simulations

Boundary Conditions

Lower Bdry:	$\frac{\partial P_{wg}}{\partial y} = \rho^w g_y = -\ 980.42$ dynes/cm^2/cm
no flow	$\frac{\partial P_{ow}}{\partial y} = (\rho^o - \rho^w) g_y = -456.87$ dynes/cm^2/cm
	$\frac{\partial \omega_1^o}{\partial y} = 0.0$
Upper Bdry:	$\frac{\partial P_{wg}}{\partial y} = -980.42$ dynes/cm^2/cm
no flow	$\frac{\partial P_{ow}}{\partial y} = -456.87$ dynes/cm^2/cm
	$\frac{\partial \omega_1^o}{\partial y} = 0.0$
at nodes 60,61:	$P_{ow} = 5.0 \times 10^4$ dynes/cm^2
	$\omega_1^o = 0.5$
	$P_{wg_{60}} = 6.5311 \times 10^4$ dynes/cm^2
	$P_{wg_{61}} = 6.6669 \times 10^4$ dynes/cm^2
Left Bdry:	$P_{wg} = \text{const}$ } hydrostatic distribution
	$P_{ow} = \text{const} = -P_{wg}$ } hydrostatic distribution
	$\frac{\partial \omega_1^o}{\partial x} = 0.0$
Right Bdry:	$P_{wg} = \text{const}$ } hydrostatic distribution + pressure head increment of 10 cm
	$P_{ow} = \text{const} = -P_{wg}$ } hydrostatic distribution + pressure head increment of 10 cm
	$\frac{\partial \omega_1^o}{\partial x} = 0.0$
Initial Conditions:	P_{wg} = hydrostatic distribution (vertical)
	$P_{ow} = -P_{wg}$ $\omega_1^o = 0.99999$

TABLE 5.3

Parameters Used in 2-D Simulations

Parameter	Value	Units	Reference Equation
Water:			
M_w	18.02		
μ_w	1.0019×10^{-2}	poise	
β_w	4.531×10^{-11}	$cm^2/dyne$	(3.22)
ρ^{wb}, P^{wb}	$0.9997964, 1.0133 \times 10^6$	g/cm^3, $dyne/cm^2$	(3.24)
TCE:			
M_1	131.4		
μ_1	5.8×10^{-1}	poise	(3.21)
β_{o1}	~ 0.0		(3.23)
ρ^{o1b}, P^{o1b}	$1.4657, 1.0133 \times 10^6$	g/m^3, $dyne/cm^2$	(3.23)
Air:			
M_A	28.97		(3.29)
z, P^g	$1.0, 1.0133 \times 10^6$	-, $dyne/cm^2$	(3.28)
Matrix:			
ε_i	0.36		
α	2.0×10^{-10}	$cm^2/dyne$	(2.14)
$k_x = k_y$	5.8231×10^{-7}	cm^2	(2.25)
Residual Saturation:			
s_{wir}	0.306		(3.30)
s_{om}	0.17		(3.30)
Dispersion/Diffusion:			
D^{mg}	0.039	cm^2/sec	(5.9)
D^{mw}, a_L^w, a_T^w	8.434×10^{-6}, 0.1, 0.02	cm^2/sec, cm, cm	(5.9), (4.26)
D^{mo}, a_L^o, a_T^o	0.0, 0.0, 0.0	cm^2/sec, cm, cm	(5.9)

Table 5.3 (cont)

Parameter	Value	Units	Reference Eq.
Partition Coeff:			
K_p^w	3.018×10^{-4}		(4.28)
K_p^g	5.549×10^{2}		(4.30)
Convergence Criteria:			
ΔP_{ow}	10.0	dyne/cm^2	
ΔP_{wg}	10.0	dyne/cm^2	
$\Delta \omega_1^o$	1.0×10^{-4}		
g,T	9.80665×10^{2}, 293.15	cm/sec^2, ^{o}K	(2.25),(3.28)

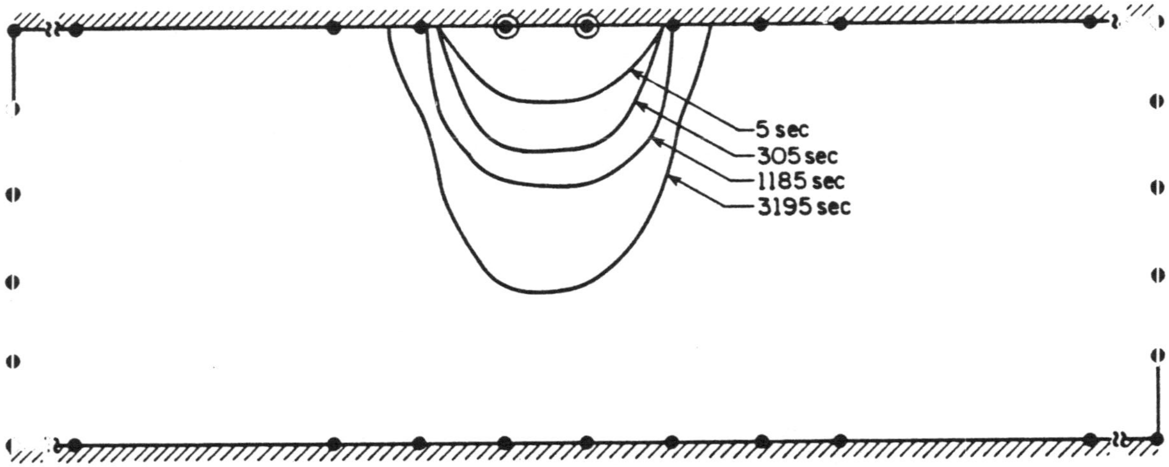

CONTOUR REPRESENTS $S_o = 0.1$

FIGURE 5.5 (a): TCE MIGRATION AS A PHASE

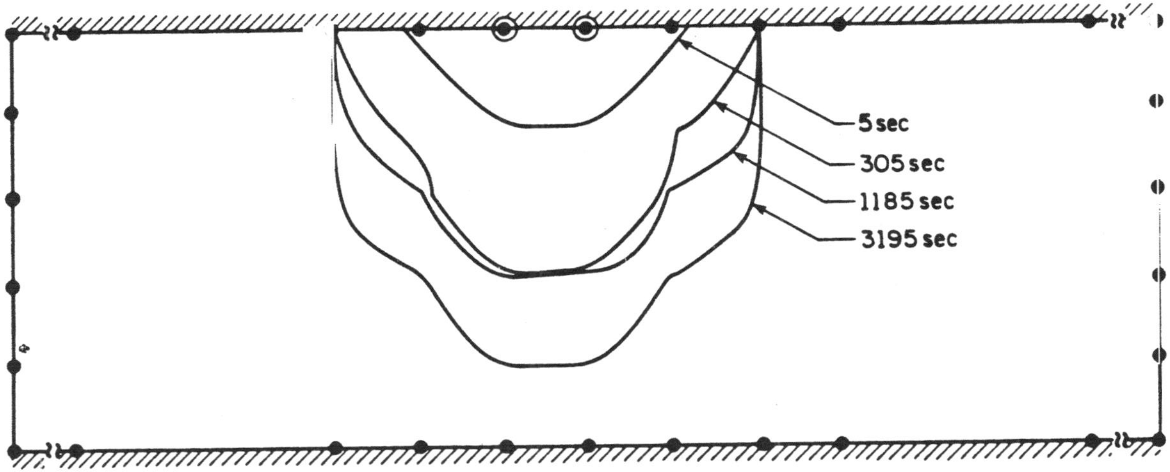

CONTOUR REPRESENTS 100 ppm TCE IN WATER PHASE

FIGURE 5.5(b): PLUME MIGRATION

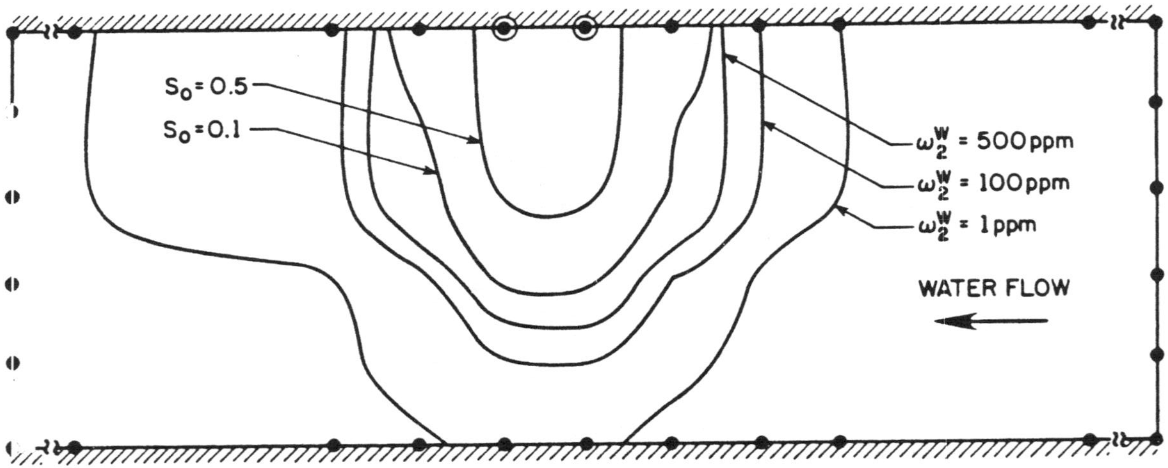

FIGURE 5.5(c): SATURATION/CONCENTRATION PLOT FOR T = 3195 SEC

the contours shown in the figures. An initial time step of 0.25 seconds was employed at the start of the simulation. Thereafter, the time step size was increased by doubling its size every other step until. a time step of 20 seconds was reached. This value was maintained throughout most of the remainder of the simulation (occasionally it was necessary to cut back the time step to 10 seconds to achieve convergence within a maximum of 15 iterations. This problem occurred at points in time when the organic saturation approached the residual value at a node.) On average, 6-7 iterations were required per time step for the convergence criteria given in Table 5.3.

TCE phase migration with time is shown in Figure 5.5(a). Here four different simulation times are represented. Each contour delineates the region within which the organic phase saturation is greater than 0.1. Although not readily apparent from the figure, the saturation distributions are skewed slightly in the direction of water flow in the aquifer; that is, saturations are slightly greater in the negative x-direction than the positive x-direction for a given horizontal distance from the source nodes.

Figure 5.5(b) depicts the corresponding migration of the dissolved organic with time. Here each contour represents a dissolved organic concentration of 100 ppm in the water phase. Because water velocities are small, the movement of the contaminant plume is governed to a great extent by the migration of the main body of the organic phase and diffusion rates of TCE in water. As with the organic phase saturation distribution, the dissolved concentrations of organic are also skewed slightly in the negative x-direction.

Figure 5.5(c) is a plot of the saturation/concentration profiles of TCE for a fixed simulation time. Contours are given for concentrations as low as 1 ppm. Notice that the plume appears to have migrated a significantly greater distance downstream than upstream. This effect, although real, is most likely exaggerated by the larger grid spacing downstream and the use of linear interpolation to obtain internode concentration values.

Another grid effect which is evident in Figures 5.5(a), 5.5(b), and 5.5(c) is the abrupt change in profile slope which occurs at points where the profile contour crosses a grid line. This effect is created by the use of linear interpolation along an element side to obtain the concentration/saturation values between neighboring nodes.

In order to assess mass conservation of the organic phase in this two-dimensional example problem, the flux of TCE into the system must be determined. Unfortunately, this flux cannot be estimated to very great accuracy. Many alternative estimating techniques are possible depending on whether one linearly interpolates flux, pressure, or saturation values, and each technique calculates a different total flux. The technique used in the model is described below. It was chosen because of its simplicity and because its predicted flux value fell in the middle of the range of the alternative estimates.

Recall that the upper-confining bed in Figure 5.4 is impermeable except for two central nodes which are saturated with TCE. Boundary conditions require that the fluxes at nodes labelled 59 and 62 in the figure be zero. Pressure gradients at nodes 60 and 61 may be approximated by one-sided difference expressions. The resulting flux of component 1 at node 60 over the time step Δt may then be written as:

$$flux_{60} = (\omega_1^o \rho^o \tau_{oy})_{60} \left[\frac{P_{og_{60}} - P_{og_{49}}}{\Delta y_{60}} - \gamma_{60}^o \right] \Delta t \qquad (5.18a)$$

A similar expression may also be written for the flux at node 61. If a linear distribution of flux between nodes is assumed, the flux midway between nodes 59 and 60 is half of $flux_{60}$, and similarly, the flux midway between 61 and 62 is half of $flux_{61}$. The total flux along the upper boundary between these two midway points is thus given by:

$$flux_{total} = (0.875 \ \Delta x_{60}) \ flux_{60} + (0.875 \ \Delta x_{61}) flux_{61} \qquad (5.18b)$$

Assuming a linear distribution of saturations between nodes, the change in mass of component 1 in the system may be approximated by:

$$\Delta \, mass = \sum_i \{ (\rho^o \varepsilon \, s_o \, \Delta x \Delta y)_i^{n+1} - (\rho^o \varepsilon \, s_o \, \Delta x \Delta y)_i^n \} \qquad (5.19)$$

Equation (5.19) is similar to the 1-D expressions for mass in equations (4.10) and (4.11). Paralleling the 1-D development, a mass balance ratio R_M, may be defined:

$$R_M = \frac{\Delta mass}{flux_{total}} \qquad (5.20)$$

This equation should be used in conjunction with equations (5.18b) and (5.19). Recall that the ratio R_M is a measure of the mass conservation properties of the simulation. It has an optimum value of 1.0.

The behavior of the ratio R_M exhibited in the 2-D simulations is illustrated by Figure 5.6. Here one sees the same type of oscillations which appeared in the 1-D TCE infiltration results (see Figure 4.16(a)). Note, however, that the variation in R_M values is amplified in Figure 5.6. The cumulative mass balance value (weighted average of R_M) is approximately 0.86 for this example.

An alternative method for calculating total mass in the system was also examined for its effects on the mass balance. In this method, pressures were linearly interpolated between nodes and a saturation profile was developed by employing the appropriate pressure/saturation relation. Mass was calculated by integrating under this saturation profile using a Gaussian quadrature procedure. Cumulative mass balance results, however, for this scheme showed no improvement over those obtained by the simpler linear interpolation of saturations.

In order to study the oscillatory mass balance behavior more closely, a 1-D example problem was constructed which mirrored the important features of the 2-D example (such as nodal spacing, initial conditions, and upper boundary conditions). The behavior of R_M for this example problem is illustrated in Figure 5.7. The effect of grid spacing on the oscillations in R_M is examined in the figure. Notice that, as the grid is successively refined,the range of oscillation decreases and the cumulative mass balance improves. Cumulative mass balance values vary from a low of 0.86 at 50 cm spacing (the same

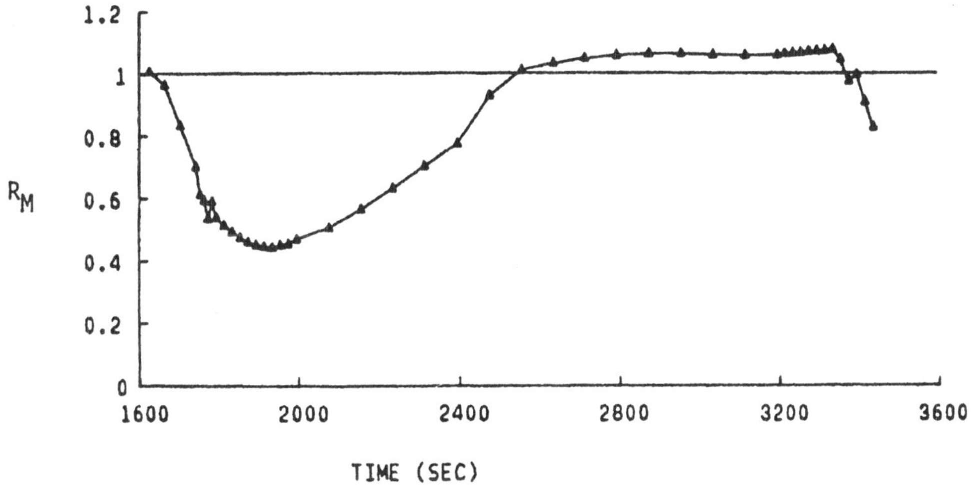

FIGURE 5.6: MASS CONSERVATION RESULTS FOR 2-D SIMULATION

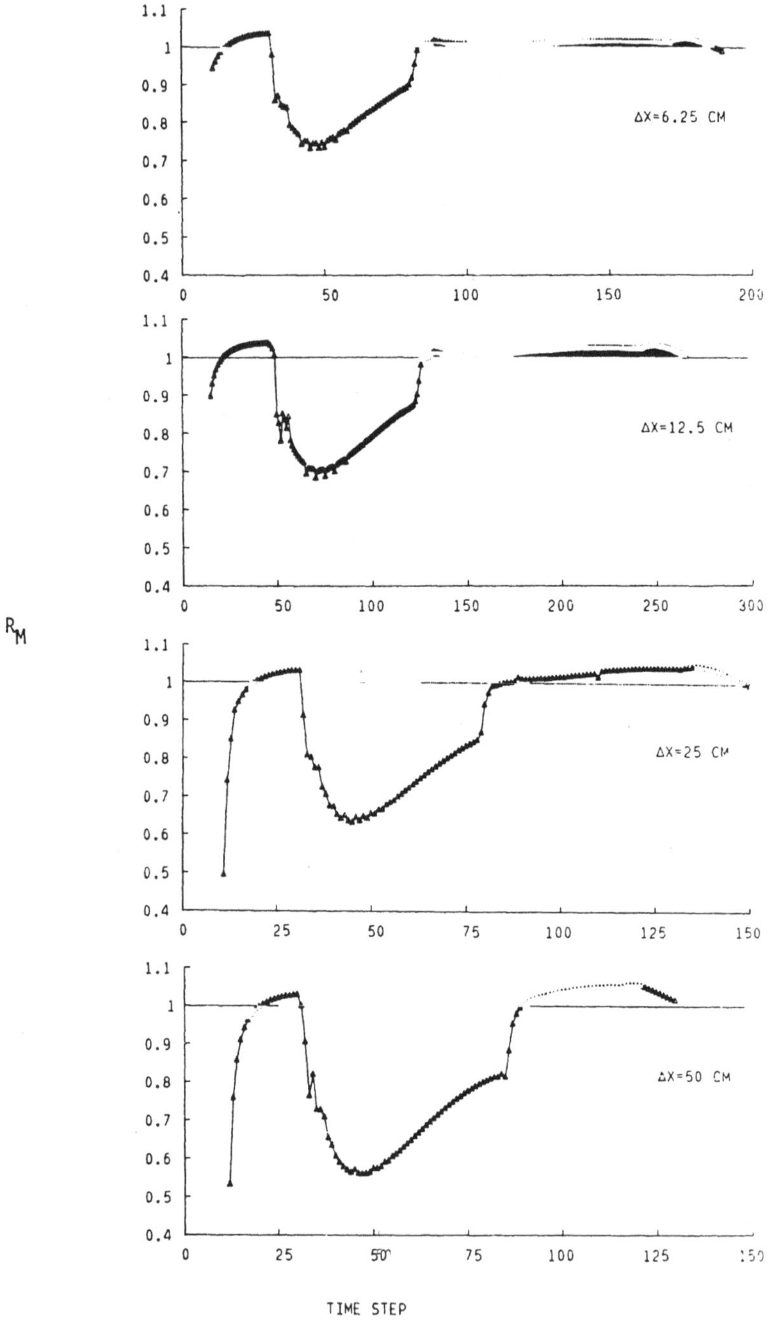

FIGURE 5.7: MASS BALANCE COMPARISON FOR SUCCESSIVE GRID REFINEMENT

spacing used in the 2-D problem) to 0.95 at a spacing of 6.25 cm. One anticipates that further grid refinement would produce values in the 0.98 range as demonstrated in Figure 4.16(a).

A comparison of the saturation profiles for various grid spacings at time $t = 260$ sec and $t = 850$ sec are shown in Figures 5.8(a) and (b). Observe that the profiles steepen with successive refinements of the grid. These figures suggest that although the coarse spacing does not permit an accurate representation of the shape of the profile, it does indicate the extent of movement of the organic. Recall that the extent of the organic phase migration determines the areal extent of the contaminant plume in the water phase. Thus, by using a coarse grid, one sacrifices some knowledge of the distribution of saturations in the multiphase region but the extent of the contaminated zone can be determined to the accuracy of a grid spacing. Note also that by keeping track of the entering flux, one has knowledge of the total mass of organic in the system. (The entering flux is not found to vary significantly with grid spacing). For many field scale problems, this information is sufficient. If more information on the saturation distribution is desired, the grid may be refined.

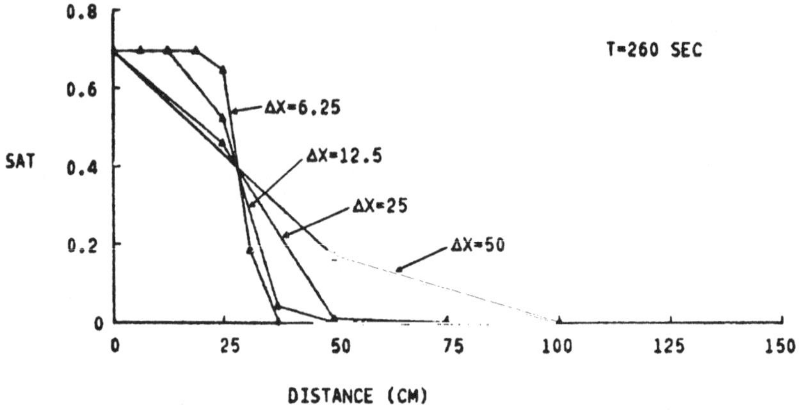

FIGURE 5.8(A): PROPAGATION OF TCE - SUCCESSIVE GRID REFINEMENT

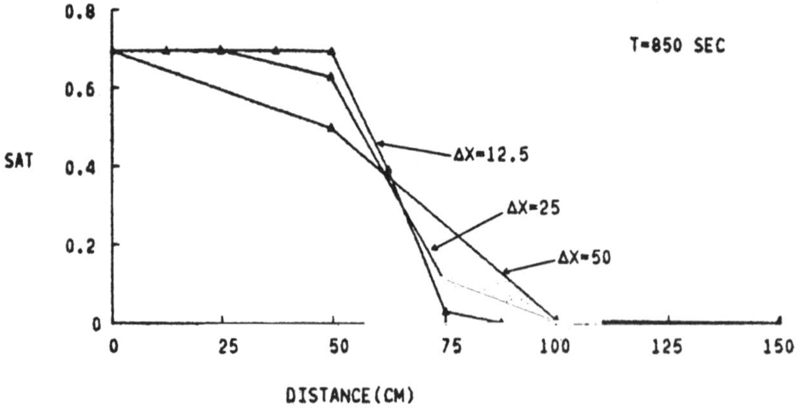

FIGURE 5.8(B): PROPAGATION OF TCE - SUCCESSIVE GRID REFINEMENT

Summary and Conclusions

Within this thesis a new multiphase approach to the modeling of organic compound migration in the subsurface has been developed. This comprehensive approach enables the modeler to track the movement of a chemical contaminant in three separate forms: as a non-aqueous phase, as a soluble component of the water phase, and as a mobile fraction of the gas phase. To implement this approach, one and two dimensional simulators were constructed. The capabilities of these simulators were then demonstrated by examples of fuel oil and TCE migration in a porous medium.

In Chapter I, a literature review disclosed the inadequacies of past modeling efforts in this area. No single model had ever been developed which was capable of examining the "near-field" contamination scenario in all its complexity. Migration of a compound in the gaseous phase was found to have been neglected entirely in the literature, and mechanisms of mass transfer had not been adequately treated. The effects of capillarity had been ignored or oversimplified. Thus, the need for a more comprehensive modeling approach was demonstrated.

Such an approach was presented in Chapter II. Here, a set of system governing equations was developed which incorporated the effects of matrix and fluid compressibility, gravity, phase composition, interphase mass exchange, capillarity, and diffusion/dispersion. No previous derivation of such a system has appeared in the literature.

In Chapter III, a numerical simulator was developed to solve this set of nonlinear partial differential equations. A fully implicit finite difference discretization method was employed. Difference operators were shown to be both consistent and conservative. Motivated by stability considerations, the scheme was made implicit in the strongly nonlinear coefficients. A Newton-Raphson iteration method was chosen to solve the resultant system of nonlinear algebraic equations. This iteration scheme was found to be convergent for all the problems examined, as long as the time step size was kept below a value dependent upon the specific boundary conditions and nodal spacing.

Chapter IV illustrated the applicability of the model to various contamination scenarios. For each simulation, convergence of the iteration scheme was demonstrated heuristically. Orders of convergence in space and time for an oil infiltration problem were found to be 2.2 and 0.98 respectively. Mass calculations for this same simulation showed that change in total mass differed less than 0.01% from the incoming flux for all time steps. Upstream weighting had little effect on the saturation profile in all situations examined and did not improve the system mass balance.

In a more highly nonlinear problem, that of TCE migration, saturation profiles were found to exhibit an "entry pressure" effect due to the nonlinearity of the pressure-saturation relation. For these simulations, maximum time step size was limited by iteration requirements at points of slowest convergence (points where the TCE relative permeability became nonzero). Time step size was also controlled by the travel time between nodes. Mass balance results for the TCE simulations exhibited a regular cyclic pattern which coincided with the arrival of the saturation front at a node. Cycle amplitude was found to diminish with reduction in nodal spacing. Experimental measurements of TCE travel time in a sand column agreed to within 2% of simulation results.

Chapter V extended the one dimensional simulator to two dimensions. A D4 node ordering scheme was successfully implemented to reduce computer storage and time requirements. The applicability of the 2-D model was demonstrated by a simulation of TCE migration in a confined aquifer system. Model determination of pollutant concentrations down to the 100 ppb range proved possible. Mass balance results exhibited the same cyclic pattern observed in the 1-D simulations. Again, grid refinement improved the mass balance at the expense of increased computation time and computer storage requirements. A coarse grid, however, was found to give an accurate measure of the extent of the contaminated zone. Such information, along with the value of the total flux of organic into the system, would be sufficient for many field problems.

Thus, a workable model has been developed which can describe much of the complexity of subsurface contaminant migration heretofore unexamined. The model can handle a wide variety of organic contaminants and porous media conditions. As more experimental data becomes available, future work may focus on the more accurate description of model parameters and the incorporation of hysteresis and adsorption effects. This model should prove a useful tool in the effort to achieve a more complete and better understanding of groundwater contamination problems.

References

Ahuja, L.R. and Swartzendruber, D., "An Improved Form of Soil-Water Diffusivity Function," Soil Sci. Soc. Amer. Proc., 36(1), 9-14, 1972.

American Petroleum Institute, "The Migration of Petroleum Products in Soil and Ground Water: Principles and Countermeasures," API Publ. 4149, Washington, D.C., Dec. 1972.

Aziz, K. and Settari, A., Petroleum Reservoir Simulation, Applied Science Publ., London, 1979.

Bartz, J. and Kaess, W., "Fuel Oil Seepage Tests in the Upper Rhine Valley: Spreading of Extra Light Fuel Oil and its Solubles Above and in Groundwater," Abh. Hess. Landesamtes Bodenforsch., 7, 1-65 (Ger) 1972.

Bastien, F., Muntzer, P., and Zilliox, L., "Pollution by Petroleum Products: Transfer of Hydrocarbons in Water and Migration of the Contaminants in the Aquifer," Prot. Eaux Souterr. Captees Aliment. Num., Commun., Colloq. Natl. 2, 1-19 (Fr), Orleans, Fr., Bur. Rech. Geol. Minieves, 1977.

Bear, J., "On the Tensor Form of Dispersion in Porous Media," J. Geophys. Res., 66(4), 1185-1197, 1961

Bear, J., Hydraulics of Groundwater, McGraw-Hill, Inc., New York, 1979.

Bear, J. and Bachmat, Y., "A Generalized Theory on Hydrodynamic Dispersion in Porous Media," IASH Symp. on Artificial Recharge and Management of Aquifers, Haifa, Israel, IASH Publ. 72, 7-16, 1967.

Bird, R.B., Stewart, W.E., and Lightfoot, E.N., Transport Phenomena, John Wiley and Sons, New York, 1960.

Blair, P.M. and Weinaug, C.F., "Solution of Two-Phase Flow Problems Using Implicit Difference Equations," Soc. Pet. Eng. J., 9, Dec., 417-424, 1969.

Boast, C.W., "Modeling the Movement of Chemicals in Soils by Water," Soil Sci., 115, 224-230, 1973.

Bresler, E., "Simultaneous Transport of Solutes and Water Under Transient Unsaturated Flow Conditions," Water Res. Res., 9(4), 975, 1973.

Brooks, R.H. and Corey, A.T., "Properties of Porous Media Affecting Fluid Flow," J. Irr. and Drain. Div. ASCE, 92 (IR2), 61-88, 1966.

Childs, E.C. and Collis-George, N., "The Permeability of Porous Materials," Proc. Roy. Soc. (London), 201A, 392-405, 1950.

Coats, K.H., "An Equation of State Compositional Model," Soc. Pet. Eng. J., Oct. , 363-376, 1980a.

Coats, K.H., "In-Situ Combustion Model," Soc. Pet. Eng. J., Dec., 533-553, 1980b.

Coats, K.H., "Reservoir Simulation: State of the Art," J. Pet. Tech., 34(8), 1982.

Coats, K.H., Nielsen, R.L., Terhune, M.H., and Weber, A.G., "Simulation of Three-Dimensional, Two-Phase Flow in Oil and Gas Reservoirs," Soc. Pet. Eng. J., 7(4), 377-388, 1967.

Corey, A.T., Rathjens, C.H., Henderson, J.H., and Wyllie, R.M.J.,"Three-Phase Relative Permeability," Trans. AIME, 207, 349-351, 1956.

Crichlow, H.B., Modern Reservoir Engineering - A Simulation Approach, Prentice-Hall, Inc., New Jersey, 1977.

Crookston, R.B., Culham, W.E., and Chen, W.H., "A Numerical Simulation Model for Thermal Recovery Processes," Soc. Pet. Eng. J., Feb., 37-58, 1979.

Dietrich, J.K. and Bondor, P.L., "Three-Phase Oil Relative Permeability Models," Soc. Pet. Eng. 6044, 51st Annual Fall Mtg., New Orleans, 1976.

Dietz, D.N., "Pollution of Permeable Strata by Oil Components," in Water Pollution by Oil,P. Hepple, ed., Applied Science Publ., Ltd., Essex, England, 127-139, 1971.

Dilling, W.L., "Interphase Transfer Processess: II. Evaporation Rates of Chloro Methanes, Ethanes, Ethylenes, Propanes, and Propylenes from Dilute Aqueous Solutions: Comparisons with Theoretical Predictions," Environ. Sci. Techn., 11(4), 405-409, 1977.

Douglas, J., Jr., "A Numerical Method for the Solution of a Parabolic System," Numer. Math., 2, 91-98, 1960.

Douglas, J., Jr., Peaceman, D.W., and Rachford, H.H., "A Method for Calculating Multi-Dimensional Immiscible Displacement," Trans. AIME., 216, 297-308, 1959.

Dracos, T., "Theoretical Considerations and Practical Implications on the Infiltration of Hydrocarbons in Aquifers," IAH International Symp. on Ground Water Pollution by Oil Hydrocarbons Proc., Prague (CS), 127-137, 1978.

Duffy, J.J., Peake, E., and Mohtadi, M.F., "Subsurface Persistence of Crude Oil Spilled on Land and its Transport in Groundwater," Oil Spill Conf. (Prev., Behavior, Control, Cleanup), New Orleans, March 8-10, API 4284, 475-478, 1977.

Duguid, J.O. and Reeves, M., "Material Transport Through Porous Media: A Finite-Element Galerkin Model," Oak Ridge National Lab., ORNL-4928, 1976.

Enfield, C.G., Carsel, R.F., Cohen, S.Z., Phan., T., and Walters, D.M., "Approximating Pollutant Transport to Ground Water," Ground Water, 20(6), 1982.

Eringen, A.C., Mechanics of Continua, Robert E. Krieger Publ. Co., Inc., Huntington, N.Y., 2nd Ed., 1980.

Freeze, R.A., "Three-Dimensional, Transient, Saturated-Unsaturated Flow in a Groundwater Basin," Water Res. Res., 7(2), 347-366, 1971.

Freeze, R.A. and Cherry, J.A., Groundwater, Prentice-Hall, Inc., Englewood Cliffs, N.J., 1979.

Fried, J.J., Muntzer, P., and Zilliox, L.,"Groundwater Pollution by Transfer of Oil Hydrocarbons," Ground Water, 17(6), 586-94, 1979.

Gallant, R.H., "Physical Properties of Hydrocarbons, Pt. 6: Chlorinated Ethylenes," Hydrocarbon Processing, 45(6), 1966.

Geraghty and Miller, Inc., Investigation of Underground Accumulation of Hydrocarbons Along Newtown Creek, Brooklyn, New York, July 1979.

Gray, W.G. and Hoffman, J.L., "A Numerical Model Study of Ground-Water Contamination from Price's Landfill, New Jersey - I. Data Base and Flow Simulation," Ground Water, 21(1), 7-14, 1983a.

Gray, W.G. and Hoffman, J.L., "A Numerical Model Study of Ground-Water Contamination from Price's Landfill, New Jersey - II. Sensitivity Analysis and Contaminant Plume Simulation," Ground Water, 21(1), 15-21, 1983b.

Gray, W.G. and Lee, P.C.Y., "On the Theorems for Local Volume Averaging of Multiphase Systems," Int. J. Multiphase Flow, 3, 333-340, 1977.

Gray, W.G. and O'Neill, K., "On the General Equations for Flow in Porous Media and their Reduction to Darcy's Law," Water Res. Res., 12(2), 148-154, 1976.

Guenther, K., "Ground-water Pollution Caused by Gasoline," Wasserwirt-Wassertech., 22(5), 158-161 (Ger), 1972.

Guerrera, A.A., "Chemical Contamination of Aquifers on Long Island, New York," J. AWWA, 73, April, 190-199, 1981.

Hassanizadeh, M. and Gray, W.G., "General Conservation Equations for Multi-Phase Systems: 1. Averaging Procedure, Adv. in Water Res., 2, Sept., 131-144, 1979a.

Hassanizadeh, M. and Gray, W.G., "General Conservation Equations for Multi-Phase Systems: 2. Mass, Momenta, Energy, and Entropy Equations," Adv. in Water Res., 2, Dec., 191-203, 1979b.

Haverkamp, R., Vauclin, M., Touma, J., Wierenga, P.J., and Vachaud, G., "A Comparison of Numerical Simulation Models for One-Dimensional Infiltration," Soil Sci. Soc. Am. J., 41(2), 285-294, 1977.

Hayduk, W. and Laudie, H., "Prediction of Diffusion Coefficients for Non-Electrolysis in Dilute Aqueous Solutions," AICHE J., 20, 611-615, 1974.

Hochmuth, D.P., "Two-Phase Flow of Immiscible Fluids in Groundwater Systems," Master's Thesis, Colorado State U., Fort Collins, CO, Fall, 1981.

Hoffman, B., "Uber die Ausbreitung geloester Kohlenwasserstoffe im Grundwasserleiter," Mitt. Inst. fur Wasserwirts. u. Landwirtz., Wasserbach, Techn. U. Hannover, V 16, Hannover (Ger), 1969.

Hoffmann, B., "Dispersion of Soluble Hydrocarbons in Ground Water Stream," in Adv. Water Poll. Res., Proc. 5th Int. Conf. 1970, Pergamon, Oxford, England, 2, HA-7b, 1-8, 1971.

Holzer, T.L., "Application of Groundwater Flow Theory to a Subsurface Oil Spill," Ground Water, 14(3), 138, 1976.

Kaess, W., "The Diffusion of Water-Soluble (Oil) Components in Groundwater," Water and Pet. Symp.,Koblenz, 1971, Gas-Wasserfach Wasser Abwasser, 113(8), 360-364 (Ger), 1972.

Kappeler, T. and Wuhrmann, K., "Microbial Degradation of the Water-Soluble Fraction of Gas Oil - I," Water Res., 12(5), 327-333, 1978.

Kilzer, L., Scheunert, I., Geyer, H., Klein, W., and Korte, F., "Laboratory Screening of the Volatilization Rates of Organic Chemicals from Water and Soil," Chemosphere, 8(10), 751-761, 1979.

Kleeberg, H.-B., "Experimentelle und theoretische Untersuchungen uber die Oelausbreitung im Boden," Mitt. Inst. fur Wasserwirts. u. Landwirts. Wasserbau, Techn. U. Hannover, V 17, Hannover (Ger), 1969.

Kolle, W. and Sontheimer, H., "The Problem of Groundwater Contamination by Petroleum Components," Brennst.-Chem., 50(4), 123-129 (Ger), 1969.

Konikow, L.F. and Bredehoeft, J.D., "Computer Model of Two-Dimensional Solute Transport and Dispersion in Ground Water," Techn. of Water Res. Invest. of the U.S. Geol. Survey, Book 7, Chapt. C2, 1978.

Lapidus, L. and Pinder, G.F., Numerical Solution of Partial Differential Equations in Science and Engineering, John Wiley and Sons, Inc., New York, 1982.

Leverett, M.C., "Capillary Behavior in Porous Solids," Trans. AIME, Petr. Div., 142, 152, 1941.

Leverett, M.C. and Lewis, W.B., "Steady Flow of Gas-Oil-Water Mixtures Through Unconsolidated Sands," Trans. AIME, 142, 107, 1941.

Lin, C., Private Communication, 1982.

Lin, C., Pinder, G.F., and Wood, E.F., "Water and Trichloroethylene as Immiscible Fluids in Porous Media," Water Resources Prog. Rept. 83-WR-2, Princeton University, Oct., 1982.

Lindorff, D.E., "Ground-Water Pollution - A Status Report," Ground Water, 17(1), 9-12, 1979.

Lohrenz, J., Bray, B.G., and Clark, C.R., "Calculating Viscosities of Reservoir Fluids from their Compositions," J. Pet. Techn., Oct., 1171-1176, 1964.

Lyman, W.J., Reehl, W.F., and Rosenblatt, D.H., Handbook of Chemical Property Estimation Methods, Environmental Behavior of Organic Compounds, McGraw-Hill, N.Y., 1982.

McKee, J.E., Laverty, F.B., Hertel, R.M., "Gasoline in Groundwater," Water Poll. Cont. Fed. J., 44(2), 293-302, 1972.

Mercer, J.W. and Faust, C.R., "The Application of Finite-Element Techniques to Immiscible Flow in Porous Media," in Finite Elements in Water Resources, Proceedings 1st International Conference, Pentech Press, London, 1.21-1.57, 1977.

Mualem, Y., "A New Model for Predicting the Hydraulic Conductivity of Unsaturated Porous Media," Water Res. Res., 12(3), 513-522, 1976.

Mull, R., "Modellmaessige Beschreibung der Ausbreitung von Mineraloelprodukten in Boden," Mitt. Inst. fur Wasserwirts. u. Landwirts. Wasserbau, Techn. U. Hannover, V 16, Hannover (Ger), 1969.

Mull, R., "Migration of Oil Products in the Subsoil with Regard to Ground Water Pollution by Oil, in Adv. Water Poll. Res., Proc. 5th Int. Conf. 1970, Pergamon, Oxford, Engl., 2, HA-7a, 1-8, 1971.

Mull, R., "Calculations and Experimental Investigations of the Migration of Hydrocarbons in Natural Soils," IAH Int. Symp. on Groundwater Poll. by Oil Hydrocarbons Proc., Prague, 167-181, 1978.

Naar, J., Wygal, R.J., and Henderson, J.H., "Imbibition Relative Permeability in Unconsolidated Porous Media," Soc. Pet. Eng. J., 2(1), 13-17, 1962.

Nathwani, J.S. and Phillips, C.R., "Adsorption-Desorption of Selected Hydrocarbons in Crude Oil on Soils," Chemosphere, 6(4), 157-162, 1977.

Nikolaevskii, V.N., "Convective Diffusion in Porous Media," PMM J. Appl. Math. Mech., 23(6), 1042-1050, 1959.

Ortega, J.M. and Rheinbolt, W.C., Iterative Solution of Nonlinear Equations in Several Variables, Academic Press, N.Y., 1970.

Osgood, J.O., "Hydrocarbon Dispersion in Ground Water: Significance and Characteristics," Ground Water, 12(6), 427-438, 1974.

Peaceman, D.W., Fundamentals of Numerical Reservoir Simulation, Elsevier Scientific Publ. Co., N.Y., 1977.

Peery, J.H. and Herron, E.H., "Three-Phase Reservoir Simulation," J. Pet. Techn., 21, 211-220, 1969.

Pelikán V., Kucera, M., and Polenka, M., "The Application of Soil Air Analysis in Order to Determine the Extent of Groundwater Contamination Due to Petroleum Products," IAH Int. Symp. on Groundwater Poll. by Oil Hydrocarbons, Proc., Prague, 73-81, 1978.

Petura, J.C., "Trichloroethylene and Methyl Chloroform in Groundwater: A Problem Assessment," J. AWWA, 73, April, 200-205, 1981.

Pickens, J.F. and Grisak, G.E., "Finite-Element Analysis of Liquid Flow, Heat Transport and Solute Transport in a Ground-Water Flow System: 1. Governing Equations and Model Formulation," Contaminant Hydrogeology Section, Hydrology Res. Div., Inland Waters Direc., Environment Canada Report, April 1979.

Pinder, G.F., "A Galerkin-Finite Element Simulation of Groundwater Contamination on Long Island, New York," Water Res. Res., 9(6), 1657-1669, 1973.

Powers, D.L., Boundary Value Problems, Academic Press, N.Y., 1972.

Price, H.S. and Coats, K.H., "Direct Methods in Reservoir Simulation," Soc. Pet. Eng. J., June, 295-308, 1974.

Raithby, G.D., "A Critical Evaluation of Upstream Differencing Applied to Problems Involving Fluid Flow," Comp. Meth. in Appl. Mech. Eng., 9, 75-103, 1976.

Rao, P.S.C. and Davidson, J.M., "Adsorption and Movement of Selected Pesticides at High Concentrations in Soils," Water Res., 13, 375-380, 1979.

Reid, R.C., Prausnitz, J.M., and Sherwood, T.K., The Properties of Gases and Liquids, McGraw-Hill, N.Y., 3rd ed., 1977.

Richtmeyer, R.D. and Morton, K.W., Difference Methods for Initial-Value Problems, John Wiley and Sons, Inc., N.Y., 1967.

Roebuck, J.F., Henderson, G.E., Douglas, J., Jr., and Ford, W.T., "The Compositional Reservoir Simulator: The Linear Model," Soc. Pet. Eng. J., March, 115-130, 1969.

Roux, P.H. and Althoff, W.F., "Investigation of Organic Contamination of Ground Water in South Brunswick Township, New Jersey," Ground Water, 18(5), 464-471, 1980.

Saffman, P.G., "Dispersion Due to Molecular Diffusion and Macroscopic Mixing in Flow Through a Network of Capillaries," J. Fluid Mech., 7(2), 194-208, 1960.

Scheidegger, A.E., "General Theory of Dispersion in Porous Media," J. Geophys. Res., 66(10), 3273-3278, 1961.

Schiegg, H.O., "Experimental Contribution to the Dynamic Capillary Fringe," Symp. on Hydrodynamic Diffusion and Dispersion in Porous Media, Pavia-Italia, 1977a.

Schiegg, H.O., "Methode zur Abschatzung der Ausbreitung von Erdolderivaten in mit Wasser und Luft Erfullten Boden," Mitt. der VAW an der ETH, Zurich, Nr. 22 (Ger), 1977b.

Schiegg, H.O.,"Grundlagen, Aufbau und Resultate von Laborexperimenten zur Erforschung von Oelausbreitung in Grundwassertraegern," Mitt. der VAW an der ETH, Zurich, Nr. 43, (Ger.), 1980.

Schwille, F., "Petroleum Contamination of the Subsoil - A Hydrological Problem," in Joint Problems of the Oil and Water Industries, P. Hepple, ed., Elsevier Publ. Co., N.Y., 23-53, 1967.

Schwille, F., "Die Migration von Mineraloel in Porosen Medien," Gwf-wasser/abwasser, 112(6), 307-311, 112(7), 331-339, 112(9), 465-472 (Ger), 1971.

Schwille, F., "Groundwater Pollution by Mineral Oil Products," in Proc. Groundwater Poll. Symp., Moscow, 1971, IAHS-AISH Publ. No. 103, 226-240, 1975.

Schwille, F. and Vorreyer, C., "Durch Mineraloel Reduzierte Grundwasser," Gwf-wasser/abwasser, 110(44), 1225-1232, 1969.

Segol, G., "A Three-Dimensional Galerkin-Finite Element Model for the Analysis of Contaminant Transport in Saturated-Unsaturated Porous Media," in Finite Elements in Water Resources, Proc. 1st Intl. Conf., Pentech Press, London, 2.123-144, 1977.

Shapiro, A.M., "An Alternative Formulation for Hydro-dynamic Dispersion in Porous Media," Proc. Euromech 143/Flow and Transport in Porous Media, Delft, Netherlands, Sept., 1981.

Sheffield, M., "Three Phase Flow Including Gravitational, Viscous and Capillary Forces," Soc. Pet. Eng. J., 9(2), 255-269, 1969.

Shutler, N.D., "Numerical, Three-Phase Simulation of the Linear Steamflood Process," Soc. Pet. Eng. J., 9(2), 232-246, 1969.

Snell, R.W., "Three-Phase Relative Permeability in Unconstituted Sand," J. Inst. Petrol.,48, 80-88, 1962.

Sowers, G.B. and Sowers, G.F., Introductory Soil Mechanics and Foundations, The Macmillan Co., N.Y., 3rd ed., 1970.

Sprenger, Fr.-D., "Untersuchungen uber die Ausbreitung von Mineraloel Produkten im Boden Durchgefuhrt an Zweidimensionalen Modellkorpern," Mitt. Inst. fur Wasserwirts. u. Landwirts. Wasserbau, Techn. U. Hannover,V17, Hannover, (Ger), 1969.

Stone, H.L., "Probability Model for Estimating Three-Phase Relative Permeability," J. Pet. Techn., 22(1), 214-218, 1970.

Stone, H.L., "Estimation of Three-Phase Relative Permeability and Residual Oil Data," J. Can. Pet. Techn., 12(4), 53-61, 1973.

"Strack Model to Establish Optimum Locations for a Drain to Collect Oil Polluted Groundwater: An Experimental and Numerical Evaluation," CONCAWE Rep. 13/77, Den Haag, 1977.

Toms, R.G., "Prevention of Oil Pollution from Minor Users," in Water Pollution by Oil, P. Hepple, ed., Applied Science Publ., Ltd., Essex, Engl., 89-95, 1971.

Trimble, R.H. and McDonald, A.E., "A Strongly Coupled, Fully Implicit Three-Dimensional, Three-Phase Well Coning Model," Soc. Pet. Eng. J., 21(4), 454-458, 1981.

Vachaud, G., "Determination of the Hydraulic Conductivity of Unsaturated Soils from an Analysis of Transient Flow Data," Water Res. Res., 3(3), 697-705, 1967.

VanDam, J., "The Migration of Hydrocarbons in a Water-Bearing Stratum," in The Joint Problems of the Oil and Water Industries, P. Hepple, ed., Elsevier Publ. Co., N.Y., 55-96, 1967.

Van der Waarden, M., Bridie, L.A.M., and Groenewoud, W.M., "Transport of Mineral Oil Components to Groundwater-I," Water Res., 5, 213-226, 1971.

Van der Waarden, M., Bridie, L.A.M., and Groenewoud, W.M., "Transport of Mineral Oil Components to Groundwater- II," Water Res. 11, 359-365, 1977.

Van Genuchten, M. Th., "Calculating the Unsaturated Hydraulic Conductivity with a New Closed-Form Analytical Model," Water Resources Prog. Rept. 78-WR-08, Princeton University, Sept. 1978.

Van Genuchten, M. Th., "A Comparison of Numerical Solutions of the One-Dimensional Unsaturated-Saturated Flow and Mass Transport Equations," Adv. in Water Res., 5, March, 47-54, 1982.

Van Genuchten, M. Th. and Wierenga, P.J., "Mass Transfer Studies in Absorbing Porous Media, I: Analytical Solutions," Soil Sci. Soc. Am. J., 40, 473-480, 1976.

Vanloocke, R., DeBorger, R., Voets, J.P., and Verstraete, W., "Soil and Groundwater Contamination by Oil Spills: Problems and Remedies," Int. J. Environ. Stud., 8(2), 99-111, 1975.

Wark, K., Thermodynamics, McGraw-Hill Book Co., N.Y., 1971.

Watson, K.K., "An Instantaneous Profile Method for Determining the Hydraulic Conductivity of Unsaturated Porous Material," Water Res. Res., 2, 709-715, 1966.

Weimar, R.A., "Prevent Groundwater Contamination Before It's Too Late," Water Wastes Eng., 17(2), 30-33; 63, 1980.

Williams, D.E. and Wilder, D.G., "Gasoline Pollution of a Ground-Water Reservoir: A Case History," Ground Water, 9(6), 50-54, 1971.

Wilson, J.T., Enfield, C.G., Dunlap, W.J., Cosby, R.L., Foster, D.A., and Baskin, L.B., "Transport and Fate of Selected Organic Pollutants in a Sandy Soil," J. Environ. Qual., 10(4), 501-506, 1981.

Whitaker, S., "Diffusion and Dispersion in Porous Media," AI CHE J., 13(3), 420-427, 1967.

Woodhull, R.S., "Groundwater Contamination in Connecticut," J. AWWA, 73, April, 188-189, 1981.

Yazicigil, H. and Sendlein, L.V.A., "Management of Ground Water Contaminated by Aromatic Hydrocarbons in the Aquifer Supplying Ames, Iowa," Ground Water, 19(6), 648-665, 1981.

Zoeteman, B.C.J., De Greef, E., and Brinkmann, F.J.J., "Persistency of Organic Contaminants in Groundwater: Lessons from Soil Pollution Incidents in the Netherlands," The Sci. of the Total Env., 21, 187-202, 1981.

Appendix A

Derivation of the Macroscopic Mass Balance Equation

Consider the four-phase system depicted in Figure A.1. It will be assumed that within each phase, the classical balance laws of continuum mechanics hold at any point. The microscopic mass balance equation for a material species i within a phase, may be written as:

$$\frac{\partial}{\partial t}(\rho\omega_i) + \underset{\sim}{\nabla}\cdot(\rho\omega_i \underset{\sim}{v}) - \underset{\sim}{\nabla}\cdot \underset{\sim}{j}_i - \rho f_i = 0 \tag{A.1}$$

where ρ is the bulk density of the phase

ρ_i is the mass concentration of species i over the total volume of the phase

$\omega_i \equiv \rho_i/\rho$ is the mass fraction of species i

$\underset{\sim}{v}$ is the local phase velocity

$\underset{\sim}{v}_i$ is the local velocity of species i

$\underset{\sim}{j}_i \equiv -\rho\omega_i(\underset{\sim}{v}_i - \underset{\sim}{v})$ is the diffusive flux of species i

f_i is the external supply of species i to the phase.

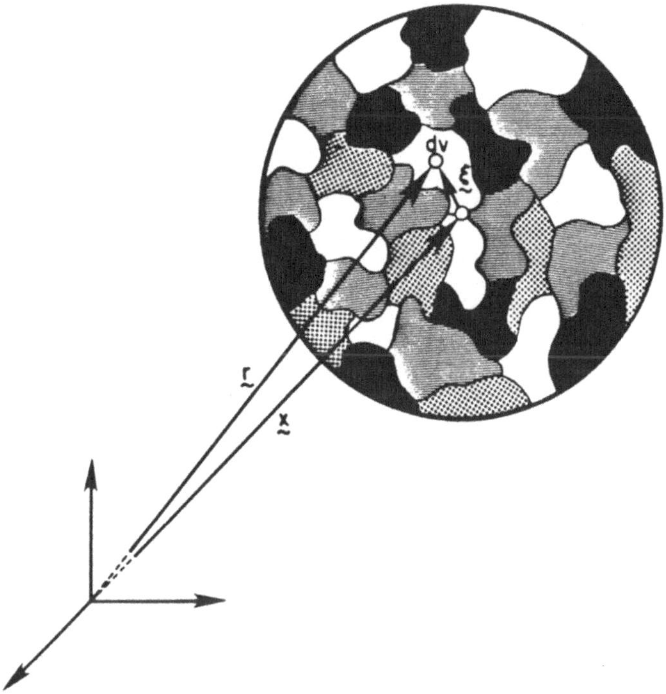

FIGURE A.1: FOUR PHASE SYSTEM AVERAGING VOLUME

(after Hassanizadeh and Gray (1979a))

Here is has been assumed that there is no net production of species i within the phase.

The balance law (A.1) is not valid on the interface between two phases α and β. At this interface, the following balance equation holds (Eringen (1980)):

$$(\rho_i(\underset{\sim}{w} - \underset{\sim}{v}) + \underset{\sim}{j}_i)\Big|_\alpha \cdot \underset{\sim}{n}^{\alpha\beta} + (\rho_i(\underset{\sim}{w} - \underset{\sim}{v}) + \underset{\sim}{j}_i)\Big|_\beta \cdot \underset{\sim}{n}^{\beta\alpha} = 0 \qquad (A.2)$$

where $\underset{\sim}{w}$ is the velocity of the interface and $\underset{\sim}{n}^{\alpha\beta}$ is the unit normal vector pointing out of the α phase into the β phase. $\Big|_\alpha$ indicates the limit of a term as the interface is approached from the α phase side. The right-hand side of (A.2) is zero because the interface does not contribute mass to the system.

To obtain a macroscopic equation from equation (A.1) by an averaging procedure, the following criteria should be satisfied (Hassanizadeh and Gray (1979 a)):

1) The characteristic length, D, of the averaging volume or area must satisfy the inequality:

$$\ell << D << L$$

where ℓ is the microscopic scale of the medium and L is the scale of gross inhomogeneities. This criteria implies that a meaningful REV (representative elementary volume) is definable.

2) The total volume of interest may contain no macroscopic surface of discontinuity.

3) Macroscopic quantities must be consistent with microscopic quantities, i.e. macroscopic quantities must account exactly for the total amount of microscopic quantities.

4) Definitions of macroscopic variables should correspond to observable and measurable functions.

5) Certain smoothness conditions must be satisfied so that the final integral equation may be localized.

Averaging operators may be defined as follows:

Volume average operator:

$$< f >_\alpha(\underset{\sim}{x},t) = \frac{1}{dV}\int_{dV} f(\underset{\sim}{x}+\underset{\sim}{\xi},t)\,\gamma(\underset{\sim}{x}+\underset{\sim}{\xi},t)\,dv(\underset{\sim}{\xi}) \tag{A.3}$$

Mass average operator:

$$\bar{f}^{\alpha}(\underset{\sim}{x},t) = \frac{1}{<\rho>_\alpha(\underset{\sim}{x},\underset{\sim}{t})dV}\int_{dV} \rho(\underset{\sim}{x}+\underset{\sim}{\xi},t)\,f(\underset{\sim}{x}+\underset{\sim}{\xi},t)\,\gamma_\alpha(\underset{\sim}{x}+\underset{\sim}{\xi},t)dv(\underset{\sim}{\xi}) \tag{A.4}$$

where dV is the averaging volume

$\underset{\sim}{\xi}$ is a microscopic spatial position vector relative to the centroid of the averaging volume (see Figure A.1)

$dv(\underset{\sim}{\xi})$ is the microscopic differential volume

$\underset{\sim}{x}$ is the position vector of the centroid of the averaging volume with respect to an inertial frame of reference

$$\gamma_\alpha(\underset{\sim}{x} + \underset{\sim}{\xi}, t) \equiv \begin{cases} 1 & \text{if } \underset{\sim}{x}+\underset{\sim}{\xi} \text{ lies within phase } \alpha \\ 0 & \text{otherwise} \end{cases}$$

γ_α is called the phase distribution function.

The averaging procedure is as follows. First, the microscopic equation is integrated over the α phase portions of the averaging volume and divided by dV. Next, the resultant equation is integrated over all the averaging volumes which encompass the space of interest. Finally, the equation is localized to achieve the macroscopic equation. In the averaging process, various theorems are employed which relate the integral of a derivative to the derivative of an integral. For a discussion of these theorems and their proofs, see Hassanizadeh and Gray (1979a) and Gray and Lee (1977).

The averaged form of the microscopic equation is:

$$\frac{\partial}{\partial t}(<\rho>_\alpha \bar{\omega}_i^\alpha) + \underset{\sim}{\nabla} \cdot (<\rho>_\alpha \bar{\underset{\sim}{v}}^\alpha \bar{\omega}_i^\alpha) - \underset{\sim}{\nabla} \cdot \underset{\sim}{J}_i^\alpha - <\rho>_\alpha \bar{f}_i^\alpha$$

$$= <\rho>_\alpha (e^\alpha(\rho\omega_i) + \hat{I}_i^\alpha) \tag{A.5}$$

where

$$e^{\alpha}(\rho\omega_i) = \frac{1}{<\rho>_{\alpha}\, dV} \sum_{\beta\neq\alpha} \int_{dA_{\alpha\beta}} \rho\omega_i(\underset{\sim}{w} - \underset{\sim}{v})\cdot\underset{\sim}{n}^{\alpha\beta}\, da \qquad (A.6)$$

and

$$\hat{I}_i^{\alpha} = \frac{1}{<\rho>_{\alpha}\, dV} \sum_{\beta\neq\alpha} \int_{dA_{\alpha\beta}} \underset{\sim}{n}^{\alpha\beta}\cdot\underset{\sim}{j}_i^{\alpha}\, da \qquad (A.7)$$

Here, $dA_{\alpha\beta}$ is the area of $\alpha\beta$-interfaces inside the averaging volume dV and da is an infinitesimal element of area in the microscopic domain. $e^{\alpha}(\rho\omega_i)$ represents the exchange of mass of species i due to phase change. $\hat{I}_i^{\alpha}$ represents the exchange of mass due to interphase diffusion. $\underset{\sim}{J}_i^{\alpha}$ is an average flux vector which represents the macroscopically non-convective flux of species i. Its existence is established by a tetrahedron argument (Hassanizadeh and Gray (1979a)). The sum of definitions (A.6) and (A.7) (the right-hand side of (A.5)) represents the total flux of species i to the α phase from all other phases. This flux is due to both convection and diffusion/dispersion and may be a positive or negative quantity. Using the definition of $\underset{\sim}{j}_i^{\alpha}$ and equations (A.6) and (A.7), the right-hand side of (A.5) may be expressed as:

$$<\rho>_{\alpha}(e^{\alpha}(\rho\omega_i) + \hat{I}_i^{\alpha}) = \frac{1}{dV} \sum_{\beta\neq\alpha} \int_{dA_{\alpha\beta}} \rho\omega_i(\underset{\sim}{w}-\underset{\sim}{v}_i)\cdot n^{\alpha\beta}\, da \qquad (A.8)$$

Equation (A.5) is subject to the following constraints:

1. The sum of the mass fractions of all species in a given phase should equal unity (from the definition of mass fraction):

$$\sum_i \bar{\omega}_i^{\alpha} = 1 \qquad (A.9)$$

2. The total flux of all species to (or from) the α phase is equal to the mass gained (or lost) by that phase (conservation principle):

$$\sum_i (e^{\alpha}(\rho\omega_i) + \hat{I}_i^{\alpha}) = e^{\alpha}(\rho) \qquad (A.10)$$

3. The mass of the total system is conserved:

$$\sum_{\alpha} e^{\alpha}(\rho) = 0 \qquad (A.11)$$

4. The mass of each species is conserved over the entire system:

$$\sum_{\alpha} (e^{\alpha}(\rho\omega_i) + \hat{I}_i^{\alpha}) = 0 \qquad (A.12)$$

Constraint (4) is derived from the averaging of (A.2) over the interfacial area and is applicable to a conservative (non-reacting) species.

Appendix B.1

Properties of the Difference Operator And its Solutions

Consider the differential operator A acting on some function u. This operator may be approximated at a point i by a finite difference operator L such that

$$A u_i = L u_i + R_i \qquad (B.1)$$

where R_i is a remainder term known as the truncation or local discretization error. Let h be the measure of the size of one mesh element. Then the difference operator L is defined as a *consistent* approximation to the differential operator A if $\|\underset{\sim}{R}\| \to 0$ as $h \to 0$, where $\|\underset{\sim}{R}\|$ is some norm of the vector $\underset{\sim}{R}$ containing elements R_i.

The remainder term R_i is most often found by manipulating Taylor series expansions for a function to produce a specific finite difference approximation plus some truncation error. The Taylor's series expansion for $u(x_i + h)$ about $u(x_i)$ may be written as:

$$u(x_i + h) = u(x_i) + h \left.\frac{\partial u}{\partial x}\right|_i + \frac{h^2}{2!} \left.\frac{\partial^2 u}{\partial x^2}\right|_i + \frac{h^3}{3!} \left.\frac{\partial^3 u}{\partial x^3}\right|_i + O\,[h^4] \qquad (B.2)$$

The quantity $O[h^4]$ is an asymptotic expression for the truncation error of this expansion. (B.2) is said to be accurate to the "order h^4." In a more rigorous sense, the notation $f = O[h^n]$ means that a positive number, B, exists independent of h such that:

$$\|f\| \leq B|h^n| \qquad \text{for all } h.$$

Now consider the error in the approximate solution to the differential equation. This error at a point i is given by:

$$e_i = u_i - \hat{u}_i \qquad (B.3)$$

where u_i is the true solution and $\hat{u}_i$ is the approximate solution. Then a solution to a finite difference equation is said to *converge* to the solution of a differential equation if $\|e\| \to 0$ as $h \to 0$. Note that consistency is a property of the difference operator while convergence is a property of the solution.

Another property of the difference operator is known as stability. A numerical scheme is called *stable* if any errors introduced into the computation (either by round-off or discretization) do not amplify during subsequent computations. The stability of a specific numerical scheme may be investigated by numerical experimentation (heuristic stability analysis), Fourier analysis (Von Neumann stability analysis), or matrix stability analysis. The latter two methods may be applied only to linear, constant coefficient finite difference approximations. For the examination of the stability of nonlinear difference equations, local linearization is necessary. See Lapidus and Pinder (1982) for a discussion of these stability analysis techniques.

The relationship between consistency, stability, and convergence for linear difference equations is expressed by Lax's Equivalence Theorem:

> Given a properly posed initial boundary value problem and a finite difference approximation to this problem which is consistent, then stability is the necessary and sufficient condition for convergence.

For a more mathematical and detailed discussion of the above concepts and of what constitutes a "properly posed" problem, see Richtmyer and Morton (1967).

Appendix B.2

Analysis of Truncation Terms

The value of a function f at the node $i+1$ (see Figure 3.1) may be expressed in terms of the function's properties at node i by use of a Taylor's series expansion (B.2):

$$f_{i+1} = f_i + \Delta x_+ \left.\frac{\partial f}{\partial x}\right|_i + \frac{(\Delta x_+)^2}{2} \left.\frac{\partial^2 f}{\partial x^2}\right|_i + \frac{(\Delta x_+)^3}{3!} \left.\frac{\partial^3 f}{\partial x^3}\right|_i + O(\Delta x_+)^4 \tag{B.4a}$$

Similarly, the value of f at $i-1$ may be written as:

$$f_{i-1} = f_i - \Delta x_- \left.\frac{\partial f}{\partial x}\right|_i + \frac{(\Delta x_-)^2}{2} \left.\frac{\partial^2 f}{\partial x^2}\right|_i - \frac{(\Delta x_-)^3}{3!} \left.\frac{\partial^3 f}{\partial x^3}\right|_i + O(\Delta x_-)^4 \tag{B.4b}$$

Subtracting (B.4b) from (B.4a) yields the expression:

$$\frac{f_{i+1} - f_{i-1}}{\Delta x_+ - \Delta x_-} = \left.\frac{\partial f}{\partial x}\right|_i + \left.\frac{\partial^2 f}{\partial x^2}\right|_i \frac{(\Delta x_+)^2 - (\Delta x_-)^2}{2(\Delta x_+ + \Delta x_-)} + \left.\frac{\partial^3 f}{\partial x^3}\right|_i \frac{(\Delta x_+)^3 + (\Delta x_-)^3}{3!(\Delta x_+ + \Delta x_-)} + O(\Delta x)^3 \tag{B.5}$$

If $\left.\frac{\partial f}{\partial x}\right|_i$ is approximated by the left hand side of (B.5), then the leading truncation term is of order Δx. Note that for the case $\Delta x_+ = \Delta x_-$ (equal nodal spacing), this error term vanishes and the approximation becomes accurate to $(\Delta x)^2$.

Now consider the Taylor's series expansion of a function at the spatial location $i + \frac{1}{2}$:

$$f_{i+\frac{1}{2}} = f_i + \frac{\Delta x_+}{2} f_i' + \frac{(\frac{\Delta x_+}{2})^2}{2} f_i'' + O(\Delta x_+)^3 \qquad \text{(B.6a)}$$

Here, partial derivatives have been represented by prime notation for convenience. An analogous expression for $f_{i-\frac{1}{2}}$ is easily obtained:

$$f_{i-\frac{1}{2}} = f_i - \frac{\Delta x_-}{2} f_i' + \frac{(\frac{\Delta x_-}{2})^2}{2} f_i'' + O(\Delta x_-)^3 \qquad \text{(B.6b)}$$

Subtraction of (B.6b) from (B.6a) gives an alternative difference expression for the derivative of f at i:

$$f_i' = \frac{f_{i+\frac{1}{2}} - f_{i-\frac{1}{2}}}{\frac{1}{2}(\Delta x_+ + \Delta x_-)} + \frac{(\frac{\Delta x_-}{2})^2 - (\frac{\Delta x_+}{2})^2}{(\Delta x_+ + \Delta x_-)} f_i'' + O(\Delta x)^3 \qquad \text{(B.7)}$$

Expressions for $f_{i+\frac{1}{2}}$ and $f_{i-\frac{1}{2}}$ are also readily obtained from Taylor's expansions:

$$f_{i+\frac{1}{2}} = \frac{1}{2}(f_{i+1} + f_i) - \frac{1}{2}\left(\frac{\Delta x_+}{2}\right)^2 f''_{i+\frac{1}{2}} + 0(\Delta x_+)^3 \qquad \text{(B.8a)}$$

$$f_{i-\frac{1}{2}} = \frac{1}{2}(f_i + f_{i-1}) - \frac{1}{2}\left(\frac{\Delta x_-}{2}\right)^2 f''_{i-\frac{1}{2}} + 0(\Delta x_-)^3 \qquad \text{(B.8b)}$$

Using the above expansions, a difference analogue to the differential operator $\frac{\partial}{\partial x}(\Omega \frac{\partial f}{\partial x})\Big|_i$ may be developed. From (B.7), this differential operator may be approximated as:

$$\frac{\partial}{\partial x}\left(\Omega \frac{\partial f}{\partial x}\right)\Bigg|_i = \frac{\left(\Omega \frac{\partial f}{\partial x}\right)_{i+\frac{1}{2}} - \left(\Omega \frac{\partial f}{\partial x}\right)_{i-\frac{1}{2}}}{\frac{1}{2}(\Delta x_+ + \Delta x_-)} + \frac{\left(\frac{\Delta x_-}{2}\right)^2 - \left(\frac{\Delta x_+}{2}\right)^2}{(\Delta x_+ + \Delta x_-)}(\Omega F')''_i + 0(\Delta x)^3 \qquad \text{(B.9)}$$

The first term in the numerator of the right hand side of (B.9) will now be examined:

$$\left(\Omega \frac{\partial f}{\partial x}\right)_{i+\frac{1}{2}} = \left(\frac{1}{2}(\Omega_{i+1} + \Omega_i) - \frac{1}{2}\left(\frac{\Delta x_+}{2}\right)^2 \Omega''_{i+\frac{1}{2}} + 0(\Delta x_+)^3\right)\left(\frac{\partial f}{\partial x}\right)_{i+\frac{1}{2}} \qquad \text{(B.10)}$$

Here, (B.8a) has been employed. Two alternative ways of expressing a function's derivative at the $i+\frac{1}{2}$ level are given by:

$$\left.\frac{\partial f}{\partial x}\right|_{i+\frac{1}{2}} = \frac{f_{i+1} - f_i}{\Delta x_+} + \frac{(\frac{\Delta x_+}{2})^2}{3!} f'''_{i+\frac{1}{2}} + O(\Delta x_+)^3 \qquad (B.11)$$

and

$$\left.\frac{\partial f}{\partial x}\right|_{i+\frac{1}{2}} = f'_i + \frac{\Delta x_+}{2} f''_i + O(\Delta x_+)^2 \qquad (B.12)$$

These equations have been obtained by manipulation of Taylor's Series approximations Equation (B.10) may now be expanded further as:

$$(\Omega \frac{\partial f}{\partial x})_{i+\frac{1}{2}} = \frac{1}{2} (\Omega_i + \Omega_{i+1}) \left[\frac{f_{i+1} - f_i}{\Delta x_+} + \frac{(\frac{\Delta x_+}{2})^2}{3!} f'''_{i+\frac{1}{2}} + O(\Delta x_+)^3\right]$$

$$- \frac{1}{2} (\frac{\Delta x_+}{2})^2 \left[\Omega''_i + \frac{\Delta x_+}{2} \Omega'''_i + O(\Delta x_+)^2\right] [f'_i + \frac{\Delta x_+}{2} f''_i + O(\Delta x_+)^2]$$

(B.13)

Here, (B.12) has been applied to both $\Omega''_{i+\frac{1}{2}}$ and $f''_{i+\frac{1}{2}}$. Additional manipulation of (B.13) and expansion of $f'''_{i+\frac{1}{2}}$ about f'''_i yields the following result:

$$\left(\Omega \frac{\partial f}{\partial x}\right)_{i+\frac{1}{2}} = \frac{1}{2}(\Omega_i + \Omega_{i+1})\left(\frac{f_{i+1} - f_i}{\Delta x_+}\right)$$

$$+ \frac{\left(\frac{\Delta x_+}{2}\right)^2}{2}\left\{\frac{f_i'''}{3}\,\Omega_i - \Omega_i''\, f_i'\right\}$$

$$+ \frac{\left(\frac{\Delta x_+}{2}\right)^3}{2}\left\{\frac{f_i^{iv}\,\Omega_i}{3} - \Omega_i'''\, f_i' - f_i''\Omega_i'' + \frac{\Omega_i' f_i'''}{3}\right\}$$

(B.14)

Incorporation of (B.14) and a similar expression obtained for $\left(\Omega \frac{\partial f}{\partial x}\right)_{i-\frac{1}{2}}$ into (B.9) produces the final result:

$$\left.\frac{\partial}{\partial x}\left(\Omega \frac{\partial f}{\partial x}\right)\right|_i = \frac{2}{\Delta x_+ + \Delta x_-}\left\{\frac{1}{2}(\Omega_i + \Omega_{i+1})\,\frac{f_{i+1} - f_i}{\Delta x_+} - \frac{1}{2}(\Omega_i + \Omega_{i-1})\,\frac{f_i - f_{i-1}}{\Delta x_-}\right\}$$

$$+ \frac{(\Delta x_-)^2 - (\Delta x_+)^2}{6(\Delta x_+ + \Delta x_-)}\left(f_i'''\,\Omega_i + 3\,\Omega_i'' f_i'\right)$$

$$+ \frac{(\Delta x_+)^3 + (\Delta x_-)^3}{24(\Delta x_- + \Delta x_+)}\left(f_i^{iv}\,\Omega_i - 3\Omega_i''' f_i' - 3f_i''\Omega_i'' + \Omega_i' f_i'''\right)$$

$$+ O(\Delta x)^3 \qquad \text{(B.15)}$$

Note that the leading error term is of order Δx. This term will vanish for equal nodal spacing. Under this condition, the difference expression (first term in (B.15)) becomes second-order accurate.

Appendix B.3

The Newton-Raphson Iteration Method

Consider the nonlinear algebraic equation given by:

$$f(u) = 0 \qquad (B.16)$$

An approximation for the root $\bar{u}$ of this equation may be obtained by the following procedure. A Taylor's series expansion (see equation (B.2)) of the function f may be written about an initial guess $u^{(0)}$:

$$f(\bar{u}) = 0 = f(u^{(0)}) + \delta^{(0)} f'(u^{(0)}) + \ldots \qquad (B.17)$$

where $\delta^{(0)} = \bar{u} - u^{(0)}$ is the error term. Note that terms in $\delta^{(0)}$ greater than order 1 have been truncated. Neglecting these terms and solving for $\delta^{(0)}$ yields:

$$\delta^{(0)} = \frac{-f(u^{(0)})}{f'(u^{(0)})} \qquad (B.18)$$

An improved estimate for $\bar{u}$ may now be calculated from this error term:

$$u^{(1)} = u^{(0)} + \delta^{(0)} \tag{B.19}$$

At this point, another Taylor's series expansion may be written about the new estimate. Continuation of this process leads to the general algorithm:

$$u^{(\nu+1)} = u^{(\nu)} - \frac{f(u^{(\nu)})}{f'(u^{(\nu)})} \qquad \nu=0,1,\ldots \tag{B.20}$$

Here, (ν) refers to the iteration level. The iteration sequence given by (B.20) is known as the Newton-Raphson method.

The process (B.20) may be extended to a system of N simultaneous equations in N variables. A truncated multivariable Taylor's expression is given by:

$$f_i(\bar{u}_j) = f_i(u_j^{(0)}) + \sum_{j=1}^{N} \left(\frac{\partial f_i}{\partial u_j}\right)^{(0)} (\bar{u}_j - u_j^{(0)}) \qquad i=1,N \tag{B.21}$$

Let $\underset{\approx}{F}$ be the Jacobian of the vector function $\underset{\sim}{f}$:

$$F_{ij}^{(\nu)} \equiv \left(\frac{\partial f_i}{\partial u_j}\right)^{(\nu)} \tag{B.22}$$

and let $\underset{\sim}{\delta}$ be a correction vector:

$$\underset{\sim}{\delta}^{(\nu)} \equiv \underset{\sim}{u}^{(\nu+1)} - \underset{\sim}{u}^{(\nu)} \tag{B.23}$$

Then Newton's method for a system of equations may be simply represented as:

$$F_{ij}^{(\nu)} \delta_j^{(\nu)} = - f_i^{(\nu)} \tag{B.24}$$

By solving this matrix equation for $\delta_j^{(\nu)}$ and using equation (B.23), an updated estimate for the unknown vector, $\underset{\sim}{u}^{(\nu+1)}$, may be obtained.

Appendix C.1

Matrix Coefficients for the 1-D Model

This Appendix contains components of A and B for the mass balance equations at node i. Let $\underset{\sim}{u}$ be an ordered vector of 9 unknowns: $[P_{ow_{i-1}}, P_{wg_{i-1}}, \omega^{o}_{1_{i-1}}, P_{ow_i}, P_{wg_i}, \omega^{o}_{1_i}, P_{ow_{i+1}}, P_{wg_{i+1}}, \omega^{o}_{1_{i+1}}]$. A given mass balance equation at node i may then be written as:

$$A_{kj}u_j = B_k \qquad\qquad j = 1, 9$$

where k is the equation number (k = 1, 3). (Summation notation applies.)

Components of $\underset{\approx}{A}$ and $\underset{\sim}{B}$ for the water balance equation (equation number 1) are as follows:

$$A_{11} = A_{17} = 0$$

$$A_{12} = -\frac{1}{(\Delta x)^2_-} [\tau_{w_{i-\frac{1}{2}}}]^{n+1} - \frac{1}{\Delta x} [\tau_w^{n+1} \beta_w \gamma_w]_i$$

$$A_{13} = A_{16} = A_{19} = 0$$

$$A_{14} = \frac{1}{\Delta t} [\varepsilon \frac{\partial S_w^{n+1}}{\partial P_{ow}} + \kappa\, \alpha\, S_w]_i$$

$$A_{15} = \frac{1}{\Delta t} [\varepsilon \frac{\partial S_w^{n+1}}{\partial P_{wg}} + S_w(\alpha + \beta_w \varepsilon)]_i + \frac{1}{(\Delta x)^2_+} [\tau_{w_{i+\frac{1}{2}}}]^{n+1}$$

$$+ \frac{1}{(\Delta x)^2_-} [\tau_{w_{i-\frac{1}{2}}}]^{n+1}$$

$$A_{18} = - \frac{1}{(\Delta x)_+^2} [\tau_w]^{n+1}_{i+\frac{1}{2}} + \frac{1}{\Delta x} [\tau_w^{n+1} \beta_w \gamma_w]_i$$

$$B_1 = \frac{1}{\Delta t} [\varepsilon \frac{\partial S_w^{n+1}}{\partial P_{wg}} + S_w(\beta_w \varepsilon + \alpha)]_i \, P^n_{wg_i}$$

$$+ \frac{1}{\Delta t} [\varepsilon \frac{\partial S_w^{n+1}}{\partial P_{ow}} + \kappa \, \alpha \, S_w]_i \, P^n_{ow_i} - \frac{1}{2\Delta x} \gamma_{w_i} (\tau_{w_{i+1}} - \tau_{w_{i-1}})^{n+1}$$

Components for the species one equation (equation 2) are as follows:

$$A_{21} = (\omega_1^o \tau_o)_i^{n+1} \; [-\gamma_o \beta^P_{o_i} \frac{1}{\Delta x} + \frac{1}{4(\Delta x)^2} \beta^P_{o_i} (P^{n+1}_{og_{i+1}} - P^{n+1}_{og_{i-1}})$$

$$+ \frac{1}{4(\Delta x)^2} \beta^1_{o_i} (\omega^{o\,n+1}_{1_{i+1}} - \omega^{o\,n+1}_{1_{i-1}})] - \frac{1}{(\Delta x)_-^2} (\omega_1^o \tau_o)^{n+1}_{i-\frac{1}{2}} = A_{22}$$

$$A_{23} = - [\frac{1}{\rho^o}]_i \frac{1}{(\Delta x)_-^2} [\rho^o \varepsilon \, S_o^{n+1} D^o]_{i-\frac{1}{2}} - \gamma_{o_i} [\tau_o]^{n+1}_{i-1} \frac{1}{2\Delta x}$$

$$- \frac{1}{\Delta x} [(\omega_1^o \tau_o)^{n+1} \beta_o^1 \gamma_o]_i$$

$$A_{24} = \frac{1}{\Delta t} [(\omega_1^o S_o \varepsilon \beta_o^P) + \varepsilon(\omega_1^o \frac{\partial S_o}{\partial P_{ow}})^{n+1} + \kappa S_o \alpha \omega_1^{o\,n+1}]_i$$

$$+ \frac{1}{(\Delta x)_-^2} (\omega_1^o \tau_o)^{n+1}_{i-\frac{1}{2}} + \frac{1}{(\Delta x)_+^2} (\omega_1^o \tau_o)^{n+1}_{i+\frac{1}{2}}$$

$$A_{25} = \frac{1}{\Delta t} [(\omega_1^o S_o \varepsilon \beta_o^P) + \varepsilon(\omega_1^o \frac{\partial S_o}{\partial P_{wg}})^{n+1} + S_o \alpha \omega_1^{o^{n+1}}]_i$$

$$+ \frac{1}{(\Delta x)_-^2} (\omega_1^o \tau_o)^{n+1}_{i-\frac{1}{2}} + \frac{1}{(\Delta x)_+^2} (\omega_1^o \tau_o)^{n+1}_{i+\frac{1}{2}}$$

$$A_{26} = \frac{1}{\Delta t} [S_o \varepsilon(\omega_1^o \beta_o^1 + 1)]_i + [\frac{1}{\rho^o}]_i \left\{ \frac{1}{(\Delta x)_+^2} [\rho^o \varepsilon S_o^{n+1} D^o]_{i+\frac{1}{2}} \right.$$

$$\left. + \frac{1}{(\Delta x)_-^2} [\rho^o \varepsilon S_o^{n+1} D^o]_{i-\frac{1}{2}} \right\}$$

$$A_{27} = (\omega_1^o \tau_o)_i^{n+1} [\gamma_o \beta_{o_i}^P \frac{1}{\Delta x} - \frac{1}{4(\Delta x)^2} \beta_{o_i}^P (P_{og_{i+1}}^{n+1} - P_{og_{i-1}}^{n+1})$$

$$- \frac{1}{4(\Delta x)^2} \beta_{o_i}^1 (\omega_{1_{i+1}}^{o^{n+1}} - \omega_{1_{i-1}}^{o^{n+1}})] - \frac{1}{(\Delta x)_+^2} (\omega_1^o \tau_o)^{n+1}_{i+\frac{1}{2}} = A_{28}$$

$$A_{29} = - [\frac{1}{\rho^o}]_i \frac{1}{(\Delta x)_+^2} [\rho^o \varepsilon S_o^{n+1} D^o]_{i+\frac{1}{2}} + \gamma_{o_i} [\tau_o]_{i+1}^{n+1} \frac{1}{2\Delta x}$$

$$+ [(\omega_1^o \tau_o)^{n+1} \beta_o^1 \gamma_o]_i \frac{1}{\Delta x}$$

$$B_2 = \frac{1}{\Delta t} [(\omega_1^o S_o \varepsilon \beta_o^P) + \varepsilon(\omega_1^o \frac{\partial S_o}{\partial P_{ow}})^{n+1} + \kappa S_o \omega_1^{o^{n+1}} \alpha]_i P_{ow_i}^n$$

$$+ \frac{1}{\Delta t} [(\omega_1^o S_o \varepsilon \beta_o^P + \varepsilon(\omega_1^o \frac{\partial S_o}{\partial P_{wg}})^{n+1} + S_o \omega_1^{o^{n+1}} \alpha]_i P_{wg_i}^n$$

$$+ \frac{1}{\Delta t} [S_o \varepsilon(\omega_1^o \beta_o^1 + 1)]_i \omega_{1_i}^{o^n}$$

Components for species two equation (equation 3) are as follows:

$$A_{31} = \rho_i^o \Big\{ - \frac{1}{(\Delta x)_-^2} (\omega_2^o \tau_o)_{i-\frac{1}{2}}^{n+1} + (\omega_2^o \tau_o)_i^{n+1} [\beta_{o_i}^P \frac{1}{4(\Delta x)^2} (P_{og_{i+1}} - P_{og_{i-1}})^{n+1}$$

$$+ \beta_{o_i}^1 \frac{1}{4(\Delta x)^2} (\omega_{1_{i+1}}^o - \omega_{1_{i-1}}^o)^{n+1} - \beta_o^P \gamma_{o_i} \frac{1}{\Delta x}] \Big\}$$

$$A_{32} = A_{31} - \rho_i^w \Big\{ \beta_w \gamma_{w_i} (\omega_2^w \tau_w)_i^{n+1} \frac{1}{2\Delta x} + (\omega_2^w \tau_w)_{i-\frac{1}{2}}^{n+1} \frac{1}{(\Delta x)_-^2} \Big\}$$

$$A_{33} = - \beta_o^1 \gamma_o \rho_i^o (\omega_2^o \tau_o)_i^{n+1} \frac{1}{\Delta x} + \gamma_w \rho_i^w (K_2^{wo} \tau_w)_{i-1}^{n+1} \frac{1}{2\Delta x}$$

$$+ \gamma_o \rho_i^o (\tau_o)_{i-1}^{n+1} \frac{1}{2\Delta x} + \frac{1}{(\Delta x)_-^2} \Big\{ [\rho^o \varepsilon S_o^{n+1} D^o]_{i-\frac{1}{2}}$$

$$+ [\rho^w \varepsilon S_w^{n+1} D^w]_{i-\frac{1}{2}} [K_2^{wo}]_{i-1}^{n+1} + [(\rho^g S_g)^{n+1} \varepsilon D^g]_{i-\frac{1}{2}} [K_2^{wo} K_2^{gw}]_{i-1}^{n+1} \Big\}$$

$$A_{34} = \frac{\varepsilon}{\Delta t} \Big\{ \omega_2^o S_o \rho^o \beta_{o_i}^P + \rho_i^w [\omega_2^w \frac{\partial S_w}{\partial P_{ow}}]_i^{n+1} + \rho_i^o [\omega_2^o \frac{\partial S_o}{\partial P_{ow}}]_i^{n+1}$$

$$+ [\rho^g \omega_2^g \frac{\partial S_g}{\partial P_{ow}}]_i^{n+1} \Big\} + \kappa \alpha_i \frac{1}{\Delta t} [S_w \rho_i^w (\omega_2^w)_i^{n+1}$$

$$+ S_o \rho_i^o (\omega_2^o)_i^{n+1} + S_g (\rho^g \omega_2^g)_i^{n+1}]$$

$$+ \rho_i^o \Big\{ \frac{1}{(\Delta x)_-^2} [\omega_2^o \tau_o]_{i-\frac{1}{2}}^{n+1} + \frac{1}{(\Delta x)_+^2} [\omega_2^o \tau_o]_{i+\frac{1}{2}}^{n+1} \Big\}$$

$$A_{35} = \frac{\varepsilon}{\Delta t}\left\{\omega_2^o S_o \rho^o \beta_{o_i}^P + \rho_i^w[\omega_2^w \frac{\partial S_w}{\partial P_{wg}}]_i^{n+1} + \rho_i^o[\omega_2^o \frac{\partial S_o}{\partial P_{wg}}]_i^{n+1}\right.$$

$$\left. + \omega_2^w S_w \rho^w \beta_{w_i} + [\rho^g \omega_2^g \frac{\partial S_g}{\partial P_{wg}}]_i^{n+1}\right\}$$

$$+ \alpha_i \frac{1}{\Delta t}[S_w \rho_i^w (\omega_2^w)_i^{n+1} + S_o \rho_i^o (\omega_2^o)_i^{n+1} + S_g(\rho^g \omega_2^g)_i^{n+1}]$$

$$+ \rho_i^o\left\{\frac{1}{(\Delta x)_-^2}[\omega_2^o \tau_o]_{i-\frac{1}{2}}^{n+1} + \frac{1}{(\Delta x)_+^2}[\omega_2^o \tau_o]_{i+\frac{1}{2}}^{n+1}\right\}$$

$$+ \rho_i^w\left\{\frac{1}{(\Delta x)_-^2}[\omega_2^w \tau_w]_{i-\frac{1}{2}}^{n+1} + \frac{1}{(\Delta x)_+^2}[\omega_2^w \tau_w]_{i+\frac{1}{2}}^{n+1}\right\}$$

$$A_{36} = \frac{\varepsilon}{\Delta t}\left\{\omega_2^o S_o \beta_o^1 \rho_i^o - \rho^o S_{o_i} - \rho^w S_{w_i}(K_{2.}^{wo})_i^{n+1}\right.$$

$$\left. - S_{g_i}(\rho^g K_2^{gw} K_2^{wo})_i^{n+1}[1 + \omega_2^g \beta_g]_i\right\}$$

$$- \frac{1}{(\Delta x)_+^2}\left\{[K_2^{wo}]_i^{n+1}[\rho^w \varepsilon S_w^{n+1} D^w]_{i+\frac{1}{2}} + [\rho^o \varepsilon S_o^{n+1} D^o]_{i+\frac{1}{2}}\right.$$

$$\left. + [K_2^{wo} K_2^{gw}]_i^{n+1}[(\rho^g S_g)^{n+1} \varepsilon D^g]_{i+\frac{1}{2}}\right\}$$

$$- \frac{1}{(\Delta x)_-^2}\left\{[K_2^{wo}]_i^{n+1}[\rho^w \varepsilon S_w^{n+1} D^w]_{i-\frac{1}{2}} + [\rho^o \varepsilon S_o^{n+1} D^o]_{i-\frac{1}{2}}\right.$$

$$\left. + [K_2^{wo} K_2^{gw}]_i^{n+1}[(\rho^g S_g)^{n+1} \varepsilon D^g]_{i-\frac{1}{2}}\right\}$$

$$A_{37} = \rho_i^o \{- \frac{1}{(\Delta x)_+^2} [\omega_2^o \tau_o]_{i+\frac{1}{2}}^{n+1} - [\omega_2^o \tau_o]_i^{n+1} [\beta_{o_i}^P \frac{1}{4(\Delta x)^2} (P_{og_{i+1}} - P_{og_{i-1}})^{n+1}$$

$$+ \beta_{o_i}^1 \frac{1}{4(\Delta x)^2} (\omega_{1_{i+1}}^o - \omega_{1_{i-1}}^o)^{n+1} - \gamma_o \beta_{o_i}^P \frac{1}{\Delta x}]\}$$

$$A_{38} = A_{37} + \rho_i^w \{\gamma_w \beta_{w_i} [\omega_2^w \tau_w]_i^{n+1} \frac{1}{2\Delta x} - [\omega_2^w \tau_w]_{i+\frac{1}{2}}^{n+1} \frac{1}{(\Delta x)_+^2}\}$$

$$A_{39} = \rho^o \gamma_o \beta_{o_i}^1 (\omega_2^o \tau_o)_i^{n+1} \frac{1}{\Delta x} - \rho^w \gamma_{w_i} (K_2^{wo} \tau_w)_{i+1}^{n+1} \frac{1}{2\Delta x}$$

$$- \rho_i^o \gamma_o (\tau_o)_{i+1}^{n+1} \frac{1}{2\Delta x} + \frac{1}{(\Delta x)_+^2} \{[\rho^o \varepsilon S_o^{n+1} D^o]_{i+\frac{1}{2}}$$

$$+ [\rho^w \varepsilon S_w^{n+1} D^w]_{i+\frac{1}{2}} [K_2^{wo}]_{i+1}^{n+1} + [(\rho^g S_g)^{n+1} \varepsilon D^g]_{i+\frac{1}{2}} [K_2^{wo} K_2^{gw}]_{i+1}^{n+1}\}$$

$$B_3 = \{\frac{\varepsilon}{\Delta t} [\omega_2^o S_o \rho^o \beta_{o_i}^p + \rho_i^w (\omega_2^w \frac{\partial S_w}{\partial P_{ow}})_i^{n+1} + \rho_i^o (\omega_2^o \frac{\partial S_o}{\partial P_{ow}})_i^{n+1}$$

$$+ (\rho^g \omega_2^g \frac{\partial S_g}{\partial P_{ow}})_i^{n+1}] + \kappa \alpha_i \frac{1}{\Delta t} [S_w \rho_i^w (\omega_2^w)_i^{n+1} + S_o \rho_i^o (\omega_2^o)_i^{n+1}$$

$$+ S_g (\rho^g \omega_2^g)_i^{n+1}]\} P_{ow_i}^n$$

$$+ \{\frac{\varepsilon}{\Delta t} [\omega_2^o S_o \rho^o \beta_{o_i}^p + \rho_i^w (\omega_2^w \frac{\partial S_w}{\partial P_{wg}})_i^{n+1} + \rho_i^o (\omega_2^o \frac{\partial S_o}{\partial P_{wg}})_i^{n+1}$$

B_3 (continued)

$$+ \omega_2^w S_w \rho^w \beta_{w_i} + (\rho^g \omega_2^g \frac{\partial S_g}{\partial P_{wg}})_i^{n+1}] + \alpha_i \frac{1}{\Delta t} [S_w \rho_i^w (\omega_2^w)_i^{n+1}$$

$$+ S_o \rho_i^o (\omega_2^o)_i^{n+1} + S_g (\rho^g \omega_2^g)_i^{n+1}]\} P_{wg_i}^n$$

$$+ \frac{\varepsilon}{\Delta t} \{[\omega_2^o \rho^o S_o \beta_{o_i}^1 - \rho^o S_{o_i}] \omega_{1_i}^{o^n} - [\rho^w S_w K_{2_i}^{wo}](\omega_{1_i}^{o^n} - 1)$$

$$- \rho^w S_w (K_2^{wo})_i^{n+1} - (S_g \rho^g)_i^{n+1} [1 + \omega_2^g \beta_g][\omega_1^o K_2^{wo} K_{2_i}^{gw}$$

$$+ (K_2^{wo} K_2^{gw})_i^{n+1} - K_2^{wo} K_{2_i}^{gw}]\}$$

$$+ \rho^o \gamma_{o_i} \frac{1}{2\Delta x} [(\tau_o)_{i-1}^{n+1} - (\tau_o)_{i+1}^{n+1}] + \frac{1}{(\Delta x)_+^2} [\varepsilon (\rho^g S_g)^{n+1} D^g]_{i+\frac{1}{2}}$$

$$\cdot ([K_2^{wo} K_2^{gw}]_{i+1}^{n+1} - [K_2^{wo} K_2^{gw}]_i^{n+1})$$

$$+ \frac{1}{(\Delta x)_-^2} [\varepsilon (\rho^g S_g)^{n+1} D^g]_{i-\frac{1}{2}} ([K_2^{wo} K_2^{gw}]_{i-1}^{n+1} - [K_2^{wo} K_2^{gw}]_i^{n+1})$$

$$+ \frac{1}{(\Delta x)_+^2} [\varepsilon \rho^w S_w^{n+1} D^w]_{i+\frac{1}{2}} ([K_2^{wo}]_{i+1}^{n+1} - [K_2^{wo}]_i^{n+1})$$

$$+ \frac{1}{(\Delta x)_-^2} [\varepsilon \rho^w S_w^{n+1} D^w]_{i-\frac{1}{2}} ([K_2^{wo}]_{i-1}^{n+1} - [K_2^{wo}]_i^{n+1})$$

$$+ \rho^w \gamma_{w_i} \frac{1}{2\Delta x} [(K_2^{wo} \tau_w)_{i-1}^{n+1} - (K_2^{wo} \tau_w)_{i+1}^{n+1}]$$

All variables are evaluated at the n time level unless otherwise indicated. The subscripts i-1, i, i+1 indicate the spatial point at which the variable is evaluated. Note that these coefficients have been written for non-uniform nodal spacing. Definitions of all terms dealing with Δx are as follows (refer to Figure 3.1):

$$\Delta x = \frac{1}{2}(\Delta x_- + \Delta x_+)$$

$$(\Delta x)^2 = (\Delta x)(\Delta x)$$

$$(\Delta x)_+^2 = (\Delta x)(\Delta x_+)$$

$$(\Delta x)_-^2 = (\Delta x)(\Delta x_-)$$

For uniform nodal spacing $(\Delta x)^2 = (\Delta x)_+^2 = (\Delta x)_-^2$.

Appendix C.2

Newton-Raphson Matrix Coefficients for the 1-D Model

This appendix contains components of $\underset{\approx}{F}$, the Newton-Raphson matrix, for equations at a node i. The ordered vector of unknowns $\underset{\sim}{\delta}^{(\nu+1)}$ is equivalent to $\underset{\sim}{u}^{(\nu+1)} - \underset{\sim}{u}^{(\nu)}$, where the superscript indicates the iteration level of evaluation and $\underset{\sim}{u}$ is the vector described in Appendix C.1.

$$F_{11} = \frac{1}{(\Delta x)^2_-} \frac{\partial}{\partial P_{ow_{i-1}}} (\tau^{(\nu)}_{w_{i-\frac{1}{2}}})[P^{(\nu)}_{wg_i} - P^{(\nu)}_{wg_{i-1}}]$$

$$- \frac{1}{2\Delta x} \gamma_{w_i} \frac{\partial}{\partial P_{ow_{i-1}}} (\tau^{(\nu)}_{w_{i-1}}) + A_{11}$$

$$F_{12} = \frac{1}{(\Delta x)^2_-} \frac{\partial}{\partial P_{wg_{i-1}}} (\tau^{(\nu)}_{w_{i-\frac{1}{2}}})[P^{(\nu)}_{wg_i} - P^{(\nu)}_{wg_{i-1}}]$$

$$- \frac{1}{2\Delta x} \gamma_{w_i} \frac{\partial}{\partial P_{wg_{i-1}}} (\tau^{(\nu)}_{w_{i-1}}) + A_{12}$$

$$F_{13} = A_{13} = 0$$

$$F_{14} = \frac{\varepsilon}{\Delta t} \{\Delta P_{ow_i} \frac{\partial^2 S_w}{\partial P^2_{ow_i}} + \Delta P_{wg_i} \frac{\partial}{\partial P_{ow_i}} (\frac{\partial S_w}{\partial P_{wg}}\Big|_i)\}$$

$$+ \frac{1}{(\Delta x)^2_-} \frac{\partial}{\partial P_{ow_i}} (\tau^{(\nu)}_{w_{i-\frac{1}{2}}})[P^{(\nu)}_{wg_i} - P^{(\nu)}_{wg_{i-1}}]$$

$$+ \frac{1}{(\Delta x)_+^2} \frac{\partial}{\partial P_{ow_i}} (\tau_{w_{i+\frac{1}{2}}}^{(\nu)})[P_{wg_i}^{(\nu)} - P_{wg_{i+1}}^{(\nu)}]$$

$$+ \beta_w \gamma_{w_i} \frac{1}{\Delta x} \frac{\partial}{\partial P_{ow_i}} (\tau_{w_i}^{(\nu)})[P_{wg_{i+1}}^{(\nu)} - P_{wg_{i-1}}^{(\nu)}] + A_{14}$$

$$F_{15} = \frac{\varepsilon}{\Delta t} \{\Delta P_{ow_i} \frac{\partial}{\partial P_{wg_i}} (\frac{\partial S_w}{\partial P_{ow}} \Big|_i) + \Delta P_{wg_i} \frac{\partial^2 S_w}{\partial P_{wg_i}^2}\}$$

$$+ \frac{1}{(\Delta x)_-^2} \frac{\partial}{\partial P_{wg_i}} (\tau_{w_{i-\frac{1}{2}}}^{(\nu)})[P_{wg_i}^{(\nu)} - P_{wg_{i-1}}^{(\nu)}]$$

$$+ \frac{1}{(\Delta x)_+^2} \frac{\partial}{\partial P_{wg_i}} (\tau_{w_{i+\frac{1}{2}}}^{(\nu)})[P_{wg_i}^{(\nu)} - P_{wg_{i+1}}^{(\nu)}]$$

$$+ \beta_w \gamma_{w_i} \frac{1}{\Delta x} \frac{\partial}{\partial P_{wg_i}} (\tau_{w_i}^{(\nu)})[P_{wg_{i+1}}^{(\nu)} - P_{wg_{i-1}}^{(\nu)}] + A_{15}$$

$$F_{16} = A_{16} = 0$$

$$F_{17} = \frac{1}{(\Delta x)_+^2} \frac{\partial}{\partial P_{ow_{i+1}}} (\tau_{w_{i+\frac{1}{2}}}^{(\nu)})[P_{wg_i}^{(\nu)} - P_{wg_{i+1}}^{(\nu)}]$$

$$+ \frac{1}{2\Delta x} \gamma_{w_i} \frac{\partial}{\partial P_{ow_{i+1}}} (\tau_{w_{i+1}}^{(\nu)}) + A_{17}$$

$$F_{18} = \frac{1}{(\Delta x)_+^2} \frac{\partial}{\partial P_{wg_{i+1}}} (\tau_{w_{i+\frac{1}{2}}}^{(\nu)})[P_{wg_i}^{(\nu)} - P_{wg_{i+1}}^{(\nu)}]$$

$$+ \frac{1}{2\Delta x} \gamma_{w_i} \frac{\partial}{\partial P_{wg_{i+1}}} (\tau_{w_{i+1}}^{(\nu)}) + A_{18}$$

$$F_{19} = A_{19} = 0$$

$$F_{21} = \frac{1}{4(\Delta x)^2} [\omega_1^o \tau_o]_i^{(\nu)} \beta_{o_i}^P [P_{og_{i+1}}^{(\nu)} - P_{og_{i-1}}^{(\nu)}]$$

$$+ \frac{1}{(\Delta x)_-^2} \frac{\partial}{\partial P_{ow_{i-1}}} (\omega_1^o \tau_o)_{i-\frac{1}{2}}^{(\nu)} [P_{og_i}^{(\nu)} - P_{og_{i-1}}^{(\nu)}]$$

$$+ \frac{1}{(\Delta x)_-^2} \frac{1}{\rho_i^o} \frac{\partial}{\partial P_{ow_{i-1}}} (\rho^o \varepsilon D^o S_o^{(\nu)})_{i-\frac{1}{2}} [\omega_{1_i}^{o(\nu)} - \omega_{1_{i-1}}^{o(\nu)}]$$

$$- \frac{1}{2\Delta x} \gamma_{o_i} \frac{\partial}{\partial P_{ow_{i-1}}} (\tau_o \omega_1^o)_{i-1}^{(\nu)} + A_{21}$$

$$F_{22} = \frac{1}{4(\Delta x)^2} [\omega_1^o \tau_o]_i^{(\nu)} \beta_{o_i}^P [P_{og_{i+1}}^{(\nu)} - P_{og_{i-1}}^{(\nu)}]$$

$$+ \frac{1}{(\Delta x)_-^2} \frac{\partial}{\partial P_{wg_{i-1}}} (\omega_1^o \tau_o)_{i-\frac{1}{2}}^{(\nu)} [P_{og_i}^{(\nu)} - P_{og_{i-1}}^{(\nu)}]$$

$$+ \frac{1}{(\Delta x)_-^2} \frac{1}{\rho_i^o} \frac{\partial}{\partial P_{wg_{i-1}}} (\rho^o \varepsilon D^o S_o^{(\nu)})_{i-\frac{1}{2}} [\omega_{1_i}^{o(\nu)} - \omega_{1_{i-1}}^{o(\nu)}]$$

$$- \frac{1}{2\Delta x} \gamma_{o_i} \frac{\partial}{\partial P_{wg_{i-1}}} (\tau_o \omega_1^o)_{i-1}^{(\nu)} + A_{22}$$

$$F_{23} = \frac{1}{4(\Delta x)^2} (\omega_1^o \tau_o)_i^{(\nu)} \beta_{o_i}^1 [P_{og_{i+1}}^{(\nu)} - P_{og_{i-1}}^{(\nu)}]$$

$$+ \frac{1}{2(\Delta x)_-^2} (\tau_o)_{i-\frac{1}{2}}^{(\nu)} [P_{og_i}^{(\nu)} - P_{og_{i-1}}^{(\nu)}] + A_{23}$$

$$F_{24} = \frac{\varepsilon}{\Delta t} \omega_{1_i}^{o^{(\nu)}} \left\{ \frac{\partial^2 S_o}{\partial P_{ow_i}^2} \Delta P_{ow_i} + \frac{\partial}{\partial P_{ow_i}} \left(\frac{\partial S_o}{\partial P_{wg}} \bigg|_i \right) \Delta P_{wg_i} \right\}$$

$$- \frac{1}{4(\Delta x)^2} \beta_{o_i}^{p} \omega_{1_i}^{o^{(\nu)}} \frac{\partial}{\partial P_{ow_i}} (\tau_o)_i^{(\nu)} [P_{og_{i+1}}^{(\nu)} - P_{og_{i-1}}^{(\nu)}]^2$$

$$+ \frac{1}{4(\Delta x)^2} \beta_{o_i}^{1} \omega_{1_i}^{o^{(\nu)}} \frac{\partial}{\partial P_{ow_i}} (\tau_o)_i^{(\nu)} [\omega_{1_{i+1}}^{o^{(\nu)}} - \omega_{1_{i-1}}^{o^{(\nu)}}]$$

$$\cdot [P_{og_{i-1}}^{(\nu)} - P_{og_{i+1}}^{(\nu)}]$$

$$+ \frac{1}{(\Delta x)_-^2} \frac{\partial}{\partial P_{ow_i}} (\omega_1^o \tau_o)_{i-\frac{1}{2}}^{(\nu)} [P_{og_i}^{(\nu)} - P_{og_{i-1}}^{(\nu)}]$$

$$+ \frac{1}{(\Delta x)_+^2} \frac{\partial}{\partial P_{ow_i}} (\omega_1^o \tau_o)_{i+\frac{1}{2}}^{(\nu)} [P_{og_i}^{(\nu)} - P_{og_{i+1}}^{(\nu)}]$$

$$+ \frac{1}{(\Delta x)_-^2} \frac{1}{\rho_i^o} \frac{\partial}{\partial P_{ow_i}} (\rho^o \varepsilon D^o S_o^{(\nu)})_{i-\frac{1}{2}} [\omega_{1_i}^{o^{(\nu)}} - \omega_{1_{i-1}}^{o^{(\nu)}}]$$

$$+ \frac{1}{(\Delta x)_+^2} \frac{1}{\rho_i^o} \frac{\partial}{\partial P_{ow_i}} (\rho^o \varepsilon D^o S_o^{(\nu)})_{i+\frac{1}{2}} [\omega_{1_i}^{o^{(\nu)}} - \omega_{1_{i+1}}^{o^{(\nu)}}]$$

$$+ \frac{1}{\Delta x} \beta_o^p \gamma_o \omega_{1_i}^{o^{(\nu)}} \frac{\partial}{\partial P_{ow_i}} (\tau_o)_i^{(\nu)} [P_{og_{i+1}}^{(\nu)} - P_{og_{i-1}}^{(\nu)}]$$

$$+ \frac{1}{\Delta x} \beta_o^1 \gamma_o \omega_{1_i}^{o^{(\nu)}} \frac{\partial}{\partial P_{ow_i}} (\tau_o)_i^{(\nu)} [\omega_{1_{i+1}}^{o^{(\nu)}} - \omega_{1_{i-1}}^{o^{(\nu)}}] + A_{24}$$

$$F_{25} = \frac{\varepsilon}{\Delta t} \omega^{o(\nu)}_{1_i} \left\{ \frac{\partial}{\partial P_{wg_i}} \left(\frac{\partial S_o}{\partial P_{ow_i}}\right) \Delta P_{ow_i} + \frac{\partial^2 S_o}{\partial P^2_{wg_i}} \Delta P_{wg_i} \right\}$$

$$- \frac{1}{4(\Delta x)^2} \beta^P_o \omega^{o(\nu)}_{1_i} \frac{\partial}{\partial P_{wg_i}} (\tau_o)^{(\nu)}_i [P^{(\nu)}_{og_{i+1}} - P^{(\nu)}_{og_{i-1}}]^2$$

$$+ \frac{1}{4(\Delta x)^2} \beta^1_o \omega^{o(\nu)}_{1_i} \frac{\partial}{\partial P_{wg_i}} (\tau_o)^{(\nu)}_i [\omega^{o(\nu)}_{1_{i+1}} - \omega^{o(\nu)}_{1_{i-1}}][P^{(\nu)}_{og_{i-1}} - P^{(\nu)}_{og_{i+1}}]$$

$$+ \frac{1}{(\Delta x)^2_-} \frac{\partial}{\partial P_{wg_i}} (\omega^o_1 \tau_o)^{(\nu)}_{i-\frac{1}{2}} [P^{(\nu)}_{og_i} - P^{(\nu)}_{og_{i-1}}]$$

$$+ \frac{1}{(\Delta x)^2_+} \frac{\partial}{\partial P_{wg_i}} (\omega^o_1 \tau_o)^{(\nu)}_{i+\frac{1}{2}} [P^{(\nu)}_{og_i} - P^{(\nu)}_{og_{i+1}}]$$

$$+ \frac{1}{(\Delta x)^2_-} \frac{1}{\rho^o_i} \frac{\partial}{\partial P_{wg_i}} (\rho^o \varepsilon D^o S^{(\nu)}_o)_{i-\frac{1}{2}} [\omega^{o(\nu)}_{1_i} - \omega^{o(\nu)}_{1_{i-1}}]$$

$$+ \frac{1}{(\Delta x)^2_+} \frac{1}{\rho^o_i} \frac{\partial}{\partial P_{wg_i}} (\rho^o \varepsilon D^o S^{(\nu)}_o)_{i+\frac{1}{2}} [\omega^{o(\nu)}_{1_i} - \omega^{o(\nu)}_{1_{i+1}}]$$

$$+ \frac{1}{\Delta x} \beta^P_o \gamma_o \omega^{o(\nu)}_{1_i} \frac{\partial}{\partial P_{wg_i}} (\tau_o)^{(\nu)}_i [P^{(\nu)}_{og_{i+1}} - P^{(\nu)}_{og_{i-1}}]$$

$$+ \frac{1}{\Delta x} \beta^1_o \gamma_o \omega^{o(\nu)}_{1_i} \frac{\partial}{\partial P_{wg_i}} (\tau_o)^{(\nu)}_i [\omega^{o(\nu)}_{1_{i+1}} - \omega^{o(\nu)}_{1_{i-1}}] + A_{25}$$

$$F_{26} = \frac{1}{\Delta x} \beta_o^P \gamma_o \tau_{o_i}^{(\nu)} [P_{og_{i+1}}^{(\nu)} - P_{og_{i-1}}^{(\nu)}] - \frac{1}{4(\Delta x)^2} \beta_o^P \tau_{o_i}^{(\nu)} [P_{og_{i+1}}^{(\nu)}$$

$$- P_{og_{i-1}}^{(\nu)}]^2 + \frac{1}{4(\Delta x)^2} \beta_o^1 \tau_{o_i}^{(\nu)} [\omega_{1_{i+1}}^{o^{(\nu)}} - \omega_{1_{i-1}}^{o^{(\nu)}}][P_{og_{i-1}}^{(\nu)}] - P_{og_{i+1}}^{(\nu)}]$$

$$+ \frac{1}{2(\Delta x)_-^2} (\tau_o)_{i-\frac{1}{2}}^{(\nu)} [P_{og_i}^{(\nu)} - P_{og_{i-1}}^{(\nu)}] + \frac{1}{2(\Delta x)_+^2} (\tau_o)_{i+\frac{1}{2}}^{(\nu)} [P_{og_i}^{(\nu)} - P_{og_{i+1}}^{(\nu)}]$$

$$+ \frac{1}{\Delta x} \beta_o^1 \gamma_o \tau_{o_i}^{(\nu)} [\omega_{1_{i+1}}^{o^{(\nu)}} - \omega_{1_{i-1}}^{o^{(\nu)}}]$$

$$+ \frac{1}{\Delta t} \left\{ [\varepsilon \frac{\partial S_o^{(\nu)}}{\partial P_{ow_i}} + \kappa S_o \alpha] \Delta P_{ow_i} + [\varepsilon \frac{\partial S_o^{(\nu)}}{\partial P_{wg_i}} + S_o \alpha] \Delta P_{wg_i} \right\} + A_{26}$$

$$F_{27} = \frac{1}{4(\Delta x)^2} [\omega_1^o \tau_o]_i^{(\nu)} \beta_{o_i}^P [P_{og_{i-1}}^{(\nu)}] - P_{og_{i+1}}^{(\nu)}]$$

$$+ \frac{1}{(\Delta x)_+^2} \frac{\partial}{\partial P_{ow_{i+1}}} (\omega_1^o \tau_o)_{i+\frac{1}{2}}^{(\nu)} [P_{og_{i-1}}^{(\nu)} \quad P_{og_{i+1}}^{(\nu)}]$$

$$+ \frac{1}{(\Delta x)_+^2} \frac{1}{\rho_i^o} \frac{\partial}{\partial P_{ow_{i+1}}} (\rho^o \varepsilon D^o S_o^{(\nu)})_{i+\frac{1}{2}} [\omega_{1_i}^{o^{(\nu)}} - \omega_{1_{i+1}}^{o^{(\nu)}}]$$

$$+ \frac{1}{2\Delta x} \gamma_{o_i} \frac{\partial}{\partial P_{ow_{i+1}}} (\tau_o \omega_1^o)_{i+1}^{(\nu)} + A_{27}$$

$$F_{28} = \frac{1}{4(\Delta x)^2} [\omega_1^o \tau_o]_i^{(\nu)} \beta_{o_i}^P [P_{og_{i-1}}^{(\nu)} - P_{og_{i+1}}^{(\nu)}]$$

$$+ \frac{1}{(\Delta x)_+^2} \frac{\partial}{\partial P_{wg_{i+1}}} (\omega_1^o \tau_o)_{i+\frac{1}{2}}^{(\nu)} [P_{og_i}^{(\nu)} - P_{og_{i+1}}^{(\nu)}]$$

F_{28} (continued)

$$+ \frac{1}{(\Delta x)_+^2} \frac{1}{\rho_i^o} \frac{\partial}{\partial P_{wg_{i+1}}} (\rho^o \varepsilon D^o S_o^{(\nu)})_{i+\frac{1}{2}} [\omega_{1_i}^{o^{(\nu)}} - \omega_{1_{i+1}}^{o^{(\nu)}}]$$

$$+ \frac{1}{2\Delta x} \gamma_{o_i} \frac{\partial}{\partial P_{wg_{i+1}}} (\tau_o \omega_1^o)_{i+1}^{(\nu)} + A_{28}$$

$$F_{29} = \frac{1}{4(\Delta x)^2} [\omega_1^o \tau_o]_i^{(\nu)} \beta_o^1 [P_{og_{i-1}}^{(\nu)} - P_{og_{i+1}}^{(\nu)}]$$

$$+ \frac{1}{2(\Delta x)_+^2} (\tau_o)_{i+\frac{1}{2}}^{(\nu)} [P_{og_i}^{(\nu)} - P_{og_{i+1}}^{(\nu)}] + A_{29}$$

$$F_{31} = \frac{\rho_i^o}{(\Delta x)_-^2} \frac{\partial}{\partial P_{ow_{i-1}}} (\omega_2^o \tau_o)_{i-\frac{1}{2}}^{(\nu)} [P_{og_i}^{(\nu)} - P_{og_{i-1}}^{(\nu)}]$$

$$+ \frac{\rho_i^w}{(\Delta x)_-^2} \frac{\partial}{\partial P_{ow_{i-1}}} (\omega_2^w \tau_w)_{i-\frac{1}{2}}^{(\nu)} [P_{wg_i}^{(\nu)} - P_{wg_{i-1}}^{(\nu)}]$$

$$+ \frac{1}{4(\Delta x)^2} \rho^o \beta_{o_i}^P (\omega_2^o \tau_o)_i^{(\nu)} [P_{og_{i+1}}^{(\nu)} - P_{og_{i-1}}^{(\nu)}]$$

$$+ \frac{1}{(\Delta x)_-^2} \frac{\partial}{\partial P_{ow_{i-1}}} (\rho^o \varepsilon D^o S_o^{(\nu)})_{i-\frac{1}{2}} [\omega_{1_{i-1}}^{o^{(\nu)}} - \omega_{1_i}^{o^{(\nu)}}]$$

$$+ \frac{1}{2\Delta x} [\omega_{1_{i-1}}^{o^{(\nu)}} - 1][\rho^o \gamma_{o_i} \frac{\partial}{\partial P_{ow_{i-1}}} (\tau_o)_{i-1}^{(\nu)} + \rho^w \gamma_{w_i} \frac{\partial}{\partial P_{ow_{i-1}}} (\tau_w K_2^{wo})_{i-1}^{(\nu)}]$$

$$+ \frac{1}{(\Delta x)_-^2} [\omega_{1_{i-1}}^{o^{(\nu)}} - 1][\frac{\partial}{\partial P_{ow_{i-1}}} (K_{2_{i-1}}^{wo} (\rho^w \varepsilon D^w S_w^{(\nu)})_{i-\frac{1}{2}})$$

F_{31} (continued)

$$+ \frac{\partial}{\partial P_{ow_{i-1}}} (K_2^{wo} K_{2_{i-1}}^{gw} (\rho^{g^{(\nu)}} \varepsilon D^g S_g^{(\nu)})_{i-\frac{1}{2}})]$$

$$+ \frac{1}{(\Delta x)_-^2} [\omega_{1_i}^{o^{(\nu)}} - 1][(K_2^{wo})_i^{(\nu)} \frac{\partial}{\partial P_{ow_{i-1}}} (\rho^w \varepsilon D^w S_w^{(\nu)})_{i-\frac{1}{2}} + (K_2^{wo} K_2^{gw})_i^{(\nu)}$$

$$\cdot \frac{\partial}{\partial P_{ow_{i-1}}} (\varepsilon D^g \rho^g S_g^{(\nu)})_{i-\frac{1}{2}}] + A_{31}$$

$$F_{32} = \frac{\rho_i^o}{(\Delta x)_-^2} \frac{\partial}{\partial P_{wg_{i-1}}} (\omega_2^o \tau_o)_{i-\frac{1}{2}}^{(\nu)} [P_{og_i}^{(\nu)} - P_{og_{i-1}}^{(\nu)}]$$

$$+ \frac{\rho_i^w}{(\Delta x)_-^2} \frac{\partial}{\partial P_{wg_{i-1}}} (\omega_2^w \tau_w)_{i-\frac{1}{2}}^{(\nu)} [P_{wg_i}^{(\nu)} - P_{wg_{i-1}}^{(\nu)}]$$

$$+ \frac{1}{4(\Delta x)^2} \rho^o \beta_{o_i}^P (\omega_2^o \tau_o)_i^{(\nu)} [P_{og_i}^{(\nu)} - P_{og_{i-1}}^{(\nu)}]$$

$$+ \frac{1}{(\Delta x)_-^2} \frac{\partial}{\partial P_{wg_{i-1}}} (\rho^o \varepsilon D^o S_o^{(\nu)})_{i-\frac{1}{2}} [\omega_{1_{i-1}}^{o^{(\nu)}} - \omega_{1_i}^{o^{(\nu)}}]$$

$$+ \frac{1}{2\Delta x} [\omega_{1_{i-1}}^{o^{(\nu)}} - 1][\rho^o \gamma_o \frac{\partial}{\partial P_{wg_{i-1}}} (\tau_o)_{i-1}^{(\nu)} + \rho^w \gamma_{w_i} \frac{\partial}{\partial P_{wg_{i-1}}} (K_2^{wo} \tau_w)_{i-1}^{(\nu)}]$$

$$+ \frac{1}{(\Delta x)_-^2} [\omega_{1_{i-1}}^{o^{(\nu)}} - 1][\frac{\partial}{\partial P_{wg_{i-1}}} (K_{2_{i-1}}^{wo} (\rho^w \varepsilon D^w S_w^{(\nu)})_{i-\frac{1}{2}})$$

$$+ \frac{\partial}{\partial P_{wg_{i-1}}} (K_2^{wo} K_{2_{i-1}}^{gw} (\rho^{g^{(\nu)}} \varepsilon D^g S_g^{(\nu)})_{i-\frac{1}{2}})]$$

F_{32} (continued)

$$- \frac{1}{(\Delta x)_-^2} [\omega_{1_i}^{o^{(\nu)}} - 1][(K_2^{wo})_i^{(\nu)} \frac{\partial}{\partial P_{wg_{i-1}}} (\rho^w \varepsilon D^w S_w^{(\nu)})_{i-\frac{1}{2}}$$

$$+ (K_2^{wo} K_2^{gw})_i^{(\nu)} \frac{\partial}{\partial P_{wg_{i-1}}} (\rho^{g^{(\nu)}} \varepsilon D^g S_g^{(\nu)})_{i-\frac{1}{2}}] + A_{32}$$

$$F_{33} = \frac{\rho_i^o}{(\Delta x)_-^2} \frac{\partial}{\partial \omega_{1_{i-1}}^o} (\omega_1^o \tau_o)_{i-\frac{1}{2}}^{(\nu)} [P_{og_i}^{(\nu)} - P_{og_{i-1}}^{(\nu)}]$$

$$+ \frac{\rho_i^o}{(\Delta x)_-^2} \frac{\partial}{\partial \omega_{1_{i-1}}^o} (\omega_2^w \tau_w)_{i-\frac{1}{2}}^{(\nu)} [P_{wg_i}^{(\nu)} - P_{wg_{i-1}}^{(\nu)}]$$

$$+ \frac{1}{4(\Delta x)^2} \rho^o \beta_{o_i}^1 (\omega_2^w \tau_o)_i^{(\nu)} [P_{og_{i+1}}^{(\nu)}] - P_{og_{i-1}}^{(\nu)}]$$

$$+ \frac{1}{(\Delta x)_-^2} [\omega_{1_{i-1}}^{o^{(\nu)}} - 1][\rho^w \varepsilon D^w S_w^{(\nu)}]_{i-\frac{1}{2}} \frac{\partial}{\partial \omega_{1_{i-1}}^o} (K_2^{wo})_{i-1}^{(\nu)}$$

$$+ \frac{1}{2\Delta x} [\omega_{1_{i-1}}^{o^{(\nu)}} - 1][\rho^w \gamma_{w_i} \frac{\partial}{\partial \omega_{1_{i-1}}^o} (K_2^{wo} \tau_w)_{i-1}^{(\nu)}]$$

$$+ \frac{1}{(\Delta x)_-^2} [\omega_{1_{i-1}}^{o^{(\nu)}} - 1][\frac{\partial}{\partial \omega_{1_{i-1}}^o} (K_2^{wo} K_{2_{i-1}}^{gw} (\rho^{g^{(\nu)}} \varepsilon D^g S_g^{(\nu)})_{i-\frac{1}{2}}]$$

$$F_{34} = \frac{\rho_i^o}{(\Delta x)_-^2} \frac{\partial}{\partial P_{ow_i}} (\omega_2^o \tau_o)_{i-\frac{1}{2}}^{(\nu)} [P_{og_i}^{(\nu)} - P_{og_{i-1}}^{(\nu)}]$$

$$+ \frac{\rho_i^w}{(\Delta x)_-^2} \frac{\partial}{\partial P_{ow_i}} (\omega_2^o \tau_o)_{i-\frac{1}{2}}^{(\nu)} [P_{og_i}^{(\nu)} - P_{wg_{i-1}}^{(\nu)}]$$

$$+ \frac{\rho_i^w}{(\Delta x)_+^2} \frac{\partial}{\partial P_{ow_i}} (\omega_2^o \tau_o)_{i+\frac{1}{2}}^{(\nu)} [P_{og_i}^{(\nu)} - P_{wg_{i+1}}^{(\nu)}]$$

$$+ \frac{\rho_i^w}{(\Delta x)_+^2} \frac{\partial}{\partial P_{ow_i}} (\omega_2^w \tau_o)_{i+\frac{1}{2}}^{(\nu)} [P_{og_i}^{(\nu)} - P_{wg_{i+1}}^{(\nu)}]$$

$$+ \frac{1}{4(\Delta x)^2} \rho^o \omega_{2_i}^{o^{(\nu)}} \frac{\partial}{\partial P_{ow_i}} (\tau_o)_i^{(\nu)} \{\beta_o^1 (\omega_{1_{i+1}}^{o^{(\nu)}} - \omega_{1_{i-1}}^{o(\nu)})(P_{og_{i-1}}^{(\nu)} - P_{og_{i+1}}^{(\nu)})$$

$$- \beta_o^P (P_{og_{i+1}}^{(\nu)} - P_{og_{i-1}}^{(\nu)})\}$$

$$+ \frac{1}{\Delta x} \rho^o \gamma_{o_i} \frac{\partial}{\partial P_{ow_i}} (\omega_2^o \tau_o)_i^{(\nu)} \{\beta_o^P [P_{og_{i+1}}^{(\nu)} - P_{og_{i-1}}^{(\nu)}] + \beta_o^1 [\omega_{1_{i+1}}^{o^{(\nu)}} - \omega_{1_{i-1}}^{o^{(\nu)}}]\}$$

$$+ \frac{1}{2\Delta x} \rho^w \beta_w \gamma_{w_i} (\omega_2^o)_i^{(\nu)} \frac{\partial}{\partial P_{ow_i}} (K_2^{wo} \tau_w)_i^{(\nu)} [P_{wg_{i+1}}^{(\nu)}] - P_{wg_{i-1}}^{(\nu)}]$$

$$+ \frac{\varepsilon}{\Delta t} (\omega_2^o)_i^{(\nu)} \{\rho_i^w \frac{\partial}{\partial P_{ow_i}} (\frac{\partial S_w}{\partial P_{ow}} K_2^{wo})_i^{(\nu)} + \rho_i^o \frac{\partial^2 S_o^{(\nu)}}{\partial P_{ow_i}^2}$$

$$+ \frac{\partial}{\partial P_{ow_i}} (\rho^g K_2^{wo} K_2^{gw} \frac{\partial S_g}{\partial P_{ow}})_i^{(\nu)}\} \Delta P_{ow_i}$$

F_{34} (continued)

$$+ \frac{\alpha}{\Delta t} (\omega_2^o)_i^{(\nu)} \{\rho^w S_{w_i}^n \frac{\partial}{\partial P_{ow_i}} (K_2^{wo})_i^{(\nu)} + S_g^n \frac{\partial}{\partial P_{ow_i}} (\rho^g K_2^{wo} K_2^{gw})_i^{(\nu)}\}$$

$$\cdot (\kappa \Delta P_{ow_i} + \Delta P_{wg_i}) + \frac{1}{(\Delta x)_-^2} \frac{\partial}{\partial P_{ow_i}} (\rho^o \varepsilon D^o S_o^{(\nu)})_{i-\frac{1}{2}} [\omega_{1_{i-1}}^{o^{(\nu)}} - \omega_{1_i}^{o^{(\nu)}}]$$

$$+ \frac{1}{(\Delta x)_+^2} \frac{\partial}{\partial P_{ow_i}} (\rho^o \varepsilon D^o S_o^{(\nu)})_{i+\frac{1}{2}} [\omega_{1_{i+1}}^{o^{(\nu)}} - \omega_{1_i}^{o^{(\nu)}}]$$

$$+ \frac{1}{(\Delta x)_-^2} [\omega_{1_{i-1}}^{o^{(\nu)}} - 1] \{\frac{\partial}{\partial P_{ow_i}} (\rho^w \varepsilon D^w S_w^{(\nu)})_{i-\frac{1}{2}} K_{2_{i-1}}^{wo^{(\nu)}}$$

$$+ \frac{\partial}{\partial P_{ow_i}} (\rho^{g^{(\nu)}} \varepsilon D^g S_g^{(\nu)})_{i-\frac{1}{2}} K_2^{wo} K_{2_{i-1}}^{gw^{(\nu)}}\}$$

$$+ \frac{1}{(\Delta x)_+^2} [\omega_{1_{i+1}}^{o^{(\nu)}} - 1] \{\frac{\partial}{\partial P_{ow_i}} (\rho^w \varepsilon D^w S_w^{(\nu)})_{i+\frac{1}{2}} K_{2_{i+1}}^{wo^{(\nu)}}$$

$$+ \frac{\partial}{\partial P_{ow_i}} (\rho^{g^{(\nu)}} \varepsilon D^g S_g^{(\nu)})_{i+\frac{1}{2}} K_2^{wo} K_{2_{i+1}}^{gw^{(\nu)}}\}$$

$$- [\omega_{1_i}^{o^{(\nu)}} - 1] \{\frac{1}{(\Delta x)_-^2} [\frac{\partial}{\partial P_{ow_i}} ((\rho^w \varepsilon D^w S_w^{(\nu)})_{i-\frac{1}{2}} K_{2_i}^{wo^{(\nu)}})$$

$$+ \frac{\partial}{\partial P_{ow_i}} ((\rho^{g^{(\nu)}} \varepsilon D^g S_g^{(\nu)})_{i-\frac{1}{2}} K_2^{wo} K_{2_i}^{gw^{(\nu)}})]$$

$$+ \frac{1}{(\Delta x)_+^2} [\frac{\partial}{\partial P_{ow_i}} ((\rho^w \varepsilon D^w S_w^{(\nu)})_{i+\frac{1}{2}} K_{2_i}^{wo^{(\nu)}})$$

$$+ \frac{\partial}{\partial P_{ow_i}} ((\rho^{g^{(\nu)}} \varepsilon D^g S_g^{(\nu)})_{i+\frac{1}{2}} K_2^{wo} K_{2_i}^{gw^{(\nu)}})]\}$$

F_{34} (continued)

$$- \frac{\varepsilon}{\Delta t} \{\rho^w S^n_{w_i} \frac{\partial}{\partial P_{ow_i}} K^{wo^{(\nu)}}_{2_i} + S^n_{g_i} \frac{\partial}{\partial P_{ow_i}} (\rho^g K^{wo}_2 K^{gw}_2)^{(\nu)}_i$$

$$+ S^n_{g_i} \omega^{g^n}_{2_i} \frac{\partial}{\partial P_{ow_i}} (\rho^g \beta_g K^{wo}_2 K^{gw}_2)^{(\nu)}_i \} \{\omega^{o^{(\nu)}}_{1_i} - 1\}$$

$$+ \frac{\varepsilon}{\Delta t} (\omega^{o^n}_{1_i} - 1)(K^{wo}_2 K^{gw}_2 S_g)^n_i [1 + (\omega^g_2 \beta_g)^n_i] \frac{\partial}{\partial P_{ow_i}} (\rho^g)^{(\nu)}_i$$

$$+ \frac{\varepsilon}{\Delta t} (\omega^o_2)^{(\nu)}_i [\rho^w_i \frac{\partial}{\partial P_{ow_i}} (\frac{\partial S_w}{\partial P_{wg}} K^{wo}_2)^{(\nu)}_i + \rho^o_i \frac{\partial}{\partial P_{ow_i}} (\frac{\partial S_o}{\partial P_{wg}})^{(\nu)}_i$$

$$+ \frac{\partial}{\partial P_{ow_i}} (\rho^g K^{wo}_2 K^{gw}_2 \frac{\partial S_g}{\partial P_{wg}})^{(\nu)}_i] \Delta P_{wg_i} + A_{34}$$

$$F_{35} = \frac{1}{(\Delta x)^2_-} \{\rho^o_i \frac{\partial}{\partial P_{wg_i}} (\omega^o_2 \tau_o)^{(\nu)}_{i-\frac{1}{2}} [P^{(\nu)}_{og_i} - P^{(\nu)}_{og_{i-1}}]$$

$$+ \rho^w_i \frac{\partial}{\partial P_{wg_i}} (\omega^w_2 \tau_w)^{(\nu)}_{i-\frac{1}{2}} [P^{(\nu)}_{wg_i} - P^{(\nu)}_{wg_{i-1}}]\}$$

$$+ \frac{1}{(\Delta x)^2_+} \{\rho^o_i \frac{\partial}{\partial P_{wg_i}} (\omega^o_2 \tau_o)^{(\nu)}_{i+\frac{1}{2}} [P^{(\nu)}_{og_i} - P^{(\nu)}_{og_{i+1}}]$$

$$+ \rho^w_i \frac{\partial}{\partial P_{wg_i}} (\omega^w_2 \tau_w)^{(\nu)}_{i+\frac{1}{2}} [P^{(\nu)}_{wg_i} - P^{(\nu)}_{wg_{i+1}}]$$

$$+ \frac{1}{4(\Delta x)^2} \rho^o \omega^{o^{(\nu)}}_{2_i} \frac{\partial}{\partial P_{wg_i}} (\tau_o)^{(\nu)}_i \{\beta^1_o (\omega^{o^{(\nu)}}_{1_{i+1}} - \omega^{o^{(\nu)}}_{1_{i-1}})(P^{(\nu)}_{og_{i-1}} - P^{(\nu)}_{og_{i+1}})$$

F_{35} (continued)

$$- \beta_o^P (P_{og_{i+1}}^{(\nu)} - P_{og_{i-1}}^{(\nu)})^2 \}$$

$$+ \frac{1}{\Delta x} \rho^o \gamma_{o_i} \frac{\partial}{\partial P_{wg_i}} (\omega_2^o \tau_o)_i^{(\nu)} \{ \beta_o^P [P_{og_{i+1}}^{(\nu)} - P_{og_{i-1}}^{(\nu)}] + \beta_o^1 [\omega_{1_{i+1}}^{o^{(\nu)}} - \omega_{1_{i-1}}^{o^{(\nu)}}]\}$$

$$+ \frac{1}{2\Delta x} \rho^w \beta_w \gamma_{w_i} (\omega_2^o)_i^{(\nu)} \frac{\partial}{\partial P_{wg_i}} (K_2^{wo} \tau_w)_i^{(\nu)} [P_{wg_{i+1}}^{(\nu)} - P_{wg_{i-1}}^{(\nu)}]$$

$$+ \frac{\varepsilon}{\Delta t} (\omega_2^o)_i^{(\nu)} \{ [\rho_i^w \frac{\partial}{\partial P_{wg_i}} (\frac{\partial S_w}{\partial P_{ow}} K_2^{wo})_i^{(\nu)} + \rho^o \frac{\partial}{\partial P_{wg_i}} (\frac{\partial S_o}{\partial P_{ow}})_i^{(\nu)}$$

$$+ \frac{\partial}{\partial P_{wg_i}} (\rho^g K_2^{wo} K_2^{gw} \frac{\partial S_g}{\partial P_{ow}})_i^{(\nu)}] \Delta P_{ow_i}$$

$$+ [\rho_i^w \frac{\partial}{\partial P_{wg_i}} (\frac{\partial S_w}{\partial P_{wg}} K_2^{wo})_i^{(\nu)} + \rho^o \frac{\partial^2 S_o^{(\nu)}}{\partial P_{wg_i}^2} + (\rho^g K_2^{wo} K_2^{gw} \frac{\partial S_g}{\partial P_{wg}})_i^{(\nu)}] \Delta P_{wg_i} \}$$

$$+ \frac{\alpha}{\Delta t} (\omega_2^o)_i^{(\nu)} \{ \rho^w S_{w_i}^n \frac{\partial}{\partial P_{wg_i}} (K_2^{wo})_i^{(\nu)} + S_g^n \frac{\partial}{\partial P_{wg_i}} (\rho^g K_2^{wo} K_2^{gw})_i^{(\nu)} \}$$

$$\cdot (\kappa \Delta P_{ow_i} + \Delta P_{wg_i}) + \frac{1}{(\Delta x)_-^2} \frac{\partial}{\partial P_{wg_i}} (\rho^o \varepsilon D^o S_o^{(\nu)})_{i-\frac{1}{2}} [\omega_{1_{i-1}}^{o^{(\nu)}} - \omega_{1_i}^{o^{(\nu)}}]$$

$$+ \frac{1}{(\Delta x)_+^2} \frac{\partial}{\partial P_{wg_i}} (\rho^o \varepsilon D^o S_o^{(\nu)})_{i+\frac{1}{2}} [\omega_{1_{i+1}}^{o^{(\nu)}} - \omega_{1_i}^{o^{(\nu)}}]$$

$$+ \frac{1}{(\Delta x)_-^2} [\omega_{1_{i-1}}^{o^{(\nu)}} - 1] \{ \frac{\partial}{\partial P_{wg_i}} (\rho^w \varepsilon D^w S_w^{(\nu)})_{i-\frac{1}{2}} K_{2_{i-1}}^{wo^{(\nu)}}$$

F_{35} (continued)

$$+ \frac{\partial}{\partial P_{wg_i}} (\rho^{g^{(\nu)}} \varepsilon D^g S_g^{(\nu)})_{i-\frac{1}{2}} K_2^{wo} K_{2_{i-1}}^{gw} \}$$

$$+ \frac{1}{(\Delta x)_+^2} [\omega_{1_{i+1}}^{o^{(\nu)}} - 1] \{ \frac{\partial}{\partial P_{wg_i}} (\rho^w \varepsilon D^w S_w^{(\nu)})_{i+\frac{1}{2}} K_{2_{i+1}}^{wo^{(\nu)}}$$

$$+ \frac{\partial}{\partial P_{wg_i}} (\rho^{g^{(\nu)}} \varepsilon D^g S_g^{(\nu)})_{i+\frac{1}{2}} K_2^{wo} K_{2_{i+1}}^{gw} \}$$

$$- [\omega_{1_i}^{o^{(\nu)}} - 1] \{ \frac{1}{(\Delta x)_-^2} [\frac{\partial}{\partial P_{wg_i}} ((\rho^w \varepsilon D^w S_w^{(\nu)})_{i-\frac{1}{2}} K_{2_i}^{wo^{(\nu)}})$$

$$+ \frac{\partial}{\partial P_{wg_i}} (\rho^{g^{(\nu)}} \varepsilon D^g S_g^{(\nu)})_{i-\frac{1}{2}} K_2^{wo} K_{2_i}^{gw^{(\nu)}})]$$

$$+ \frac{1}{(\Delta x)_+^2} [\frac{\partial}{\partial P_{wg_i}} (\rho^w \varepsilon D^w S_w^{(\nu)})_{i+\frac{1}{2}} K_{2_i}^{wo^{(\nu)}}$$

$$+ \frac{\partial}{\partial P_{wg_i}} ((\rho^{g^{(\nu)}} \varepsilon D^g S_g^{(\nu)})_{i+\frac{1}{2}} K_2^{wo} K_{2_i}^{gw^{(\nu)}})]\}$$

$$- \frac{\varepsilon}{\Delta t} \{ \rho^w S_{w_i}^n \frac{\partial}{\partial P_{wg_i}} K_{2_i}^{wo^{(\nu)}} + S_{g_i}^n \frac{\partial}{\partial P_{wg_i}} (\rho^g \beta_g K_2^{wo} K_2^{gw})_i^{(\nu)}$$

$$+ S_{g_i}^n \omega_{2_i}^{g^n} \frac{\partial}{\partial P_{wg_i}} (\rho^g \beta_g K_2^{wo} K_2^{gw})_i^{(\nu)} \} \{ \omega_{1_i}^{o^{(\nu)}} - 1 \}$$

$$+ \frac{\varepsilon}{\Delta t} (\omega_1^{o^n} - 1)(K_2^{wo} K_2^{gw} S_g)_i^n (1 + (\omega_2^g \beta_g)_i^n) \frac{\partial}{\partial P_{wg_i}} (\rho^g)_i^{(\nu)} + A_{35}$$

$$F_{36} = \frac{\rho_i^o}{(\Delta x)_-^2} \left\{ \frac{\partial}{\partial \omega_{1_i}^o} (\omega_2^o \tau_o)_{i-\frac{1}{2}}^{(\nu)} [P_{og_i}^{(\nu)} - P_{og_{i-1}}^{(\nu)}] \right\}$$

$$+ \frac{\rho_i^o}{(\Delta x)_+^2} \left\{ \frac{\partial}{\partial \omega_{1_i}^o} (\omega_2^o \tau_o)_{i+\frac{1}{2}}^{(\nu)} [P_{og_i}^{(\nu)} - P_{og_{i+1}}^{(\nu)}] \right\}$$

$$+ \frac{1}{4(\Delta x)^2} \rho^o \tau_{o_i}^{(\nu)} \{ \beta_o^P (P_{og_{i-1}}^{(\nu)} - P_{og_{i+1}}^{(\nu)})^2$$

$$+ \beta_o^1 (\omega_{1_{i-1}}^{o^{(\nu)}} - \omega_{1_{i+1}}^{o^{(\nu)}})(P_{og_{i-1}}^{(\nu)} - P_{og_{i+1}}^{(\nu)}) \}$$

$$+ \frac{1}{(\Delta x)} \rho^o \gamma_o \tau_{o_i}^{(\nu)} \{ \beta_o^P (P_{og_{i-1}}^{(\nu)} - P_{og_{i+1}}^{(\nu)}) + \beta_o^1 (\omega_{1_{i-1}}^{o^{(\nu)}} - \omega_{1_{i+1}}^{o^{(\nu)}}) \}$$

$$+ \frac{\varepsilon}{\Delta t} \{ [\rho_i^w \frac{\partial S_w^{(\nu)}}{\partial P_{ow_i}} \frac{\partial}{\partial \omega_{1_i}^o} (\omega_2^o K_2^{wo})_i^{(\nu)} - \rho_i^o \frac{\partial S_o^{(\nu)}}{\partial P_{ow_i}}$$

$$+ \frac{\partial S_g^{(\nu)}}{\partial P_{ow_i}} \frac{\partial}{\partial \omega_{1_i}^o} (\rho^g \omega_2^o K_2^{wo} K_2^{gw})_i^{(\nu)}] \Delta P_{ow_i}$$

$$+ [\rho_i^w \frac{\partial S_w^{(\nu)}}{\partial P_{wg_i}} \frac{\partial}{\partial \omega_{1_i}^o} (\omega_2^o K_?^{wo})_i^{(\nu)} - \rho_i^o \frac{\partial S_o^{(\nu)}}{\partial P_{wg_i}}$$

$$+ \frac{\partial S_g^{(\nu)}}{\partial P_{wg_i}} \frac{\partial}{\partial \omega_{1_i}^o} (\rho^g \omega_2^o K_2^{wo} K_2^{gw})_i^{(\nu)}] \Delta P_{wg_i} \}$$

$$+ \frac{\alpha}{\Delta t} [\rho^w S_{w_i}^n \frac{\partial}{\partial \omega_1^o} (\omega_2^o K_2^{wo})_i^{(\nu)} - \rho^o S_{o_i}^n + S_{g_i}^n \frac{\partial}{\partial \omega_1^o} (\rho^g \omega_2^o K_2^{wo} K_2^{gw})_i^{(\nu)}]$$

F_{36} (continued)

$$\cdot\ [\Delta P_{wg_i} + \kappa \Delta P_{ow_i}] + \frac{1}{2\Delta x} \rho^w \gamma_w \beta_w \tau_{w_i}^{(\nu)} \frac{\partial}{\partial \omega_1^o} (\omega_2^o K_2^{wo})_i^{(\nu)} [P_{wg_{i+1}}^{(\nu)} - P_{wg_{i-1}}^{(\nu)}]$$

$$+ \frac{\rho_i^w}{(\Delta x)_-^2} \frac{\partial}{\partial \omega_{1_i}^o} (\omega_2^w \tau_w)_{i-\frac{1}{2}}^{(\nu)} [P_{wg_i}^{(\nu)} - P_{wg_{i-1}}^{(\nu)}]$$

$$+ \frac{\rho_i^w}{(\Delta x)_+^2} \frac{\partial}{\partial \omega_{1_i}^o} (\omega_2^w \tau_w)_{i+\frac{1}{2}}^{(\nu)} [P_{wg_i}^{(\nu)} - P_{wg_{i+1}}^{(\nu)}]$$

$$- \frac{1}{(\Delta x)_-^2} [\omega_{1_i}^{o^{(\nu)}} - 1][(\rho^w \varepsilon D^w S_w^{(\nu)})_{i-\frac{1}{2}} \frac{\partial}{\partial \omega_{1_i}^o} (K_2^{wo})_i^{(\nu)}$$

$$+ \frac{\partial}{\partial \omega_{1_i}^o} ((\rho^{g^{(\nu)}} \varepsilon D^g S_g^{(\nu)})_{i-\frac{1}{2}} K_2^{wo} K_{2_i}^{gw^{(\nu)}})]$$

$$- \frac{1}{(\Delta x)_+^2} [\omega_{1_i}^{o^{(\nu)}} - 1][(\rho^w \varepsilon D^w S_w^{(\nu)})_{i+\frac{1}{2}} \frac{\partial}{\partial \omega_{1_i}^o} (K_2^{wo})_i^{(\nu)}$$

$$+ \frac{\partial}{\partial \omega_{1_i}^o} ((\rho^{g^{(\nu)}} \varepsilon D^g S_g^{(\nu)})_{i+\frac{1}{2}} K_2^{wo} K_{2_i}^{gw^{(\nu)}})]$$

$$+ \frac{1}{(\Delta x)_-^2} [\omega_{1_{i-1}}^{o^{(\nu)}} - 1][K_2^{wo} K_{2_{i-1}}^{gw^{(\nu)}} \frac{\partial}{\partial \omega_{1_i}^o} (\rho^{g^{(\nu)}} \varepsilon D^g S_g^{(\nu)})_{i-\frac{1}{2}}]$$

$$+ \frac{1}{(\Delta x)_+^2} [\omega_{1_{i+1}}^{o^{(\nu)}} - 1][K_2^{wo} K_{2_{i+1}}^{gw^{(\nu)}} \frac{\partial}{\partial \omega_{1_i}^o} (\rho^{g^{(\nu)}} \varepsilon D^g S_g^{(\nu)})_{i+\frac{1}{2}}]$$

$$- \frac{\varepsilon}{\Delta t} [\omega_{1_i}^{o^{(\nu)}} - 1]\{\rho^w S_{w_i}^n \frac{\partial}{\partial \omega_{1_i}^o} (K_2^{wo})_i^{(\nu)}$$

F_{36} (continued)

$$+ S^n_{g_i}(1 + \beta_g\omega^g_2)^n_i \frac{\partial}{\partial\omega^o_{1_i}} (\rho^g K^{wo}_2 K^{gw}_2)^{(\nu)}_i\}$$

$$+ \frac{\varepsilon}{\Delta t} [\omega^{o^n}_{1_i} - 1]S^n_g(K^{wo}_2 K^{gw}_2)^n_i(1 + \omega^g_2\beta_g)^n_i \frac{\partial}{\partial\omega^o_{1_i}} (\rho^g)^{(\nu)}_i + A_{36}$$

$$F_{37} = \frac{\rho^o_i}{(\Delta x)^2_+} \frac{\partial}{\partial P_{ow_{i+1}}} (\omega^o_2\tau_o)^{(\nu)}_{i+\frac{1}{2}} [P^{(\nu)}_{og_i} - P^{(\nu)}_{og_{i+1}}]$$

$$+ \frac{\rho^w_i}{(\Delta x)^2_+} \frac{\partial}{\partial P_{ow_{i+1}}} (\omega^w_2\tau_w)^{(\nu)}_{i+\frac{1}{2}} [P^{(\nu)}_{wg_i} - P^{(\nu)}_{wg_{i+1}}]$$

$$+ \frac{1}{4(\Delta x)^2} \rho^o\beta^P_{o_i} (\omega^o_2\tau_o)^{(\nu)}_i [P^{(\nu)}_{wg_{i-1}} - P^{(\nu)}_{wg_{i+1}}]$$

$$+ \frac{1}{(\Delta x)^2_+} \frac{\partial}{\partial P_{ow_{i-1}}} (\rho^o\varepsilon D^o S^{(\nu)}_o)_{i+\frac{1}{2}} [\omega^{o^{(\nu)}}_{1_{i+1}} - \omega^{o^{(\nu)}}_{1_i}]$$

$$- \frac{1}{2\Delta x} [\omega^{o^{(\nu)}}_{1_{i+1}} - 1][\rho^o\gamma_{o_i} \frac{\partial}{\partial P_{ow_{i+1}}} (\tau_o)^{(\nu)}_{i+1} + \rho^w\gamma_{w_i} \frac{\partial}{\partial P_{ow_{i+1}}} (K^{wo}_2\tau_o)^{(\nu)}_{i+1}]$$

$$+ \frac{1}{(\Delta x)^2_+} [\omega^{o^{(\nu)}}_{1_{i+1}} - 1][\frac{\partial}{\partial P_{ow_{i+1}}} (K^{wo}_{2_{i+1}} (\rho^w\varepsilon D^w S^{(\nu)}_w)_{i+\frac{1}{2}}$$

$$+ \frac{\partial}{\partial P_{ow_{i+1}}} (K^{wo}_2 K^{gw}_{2_{i+1}} (\rho^{g^{(\nu)}} \varepsilon D^g S^{(\nu)}_g)_{i+\frac{1}{2}}]$$

$$+ \frac{1}{(\Delta x)^2_+} [\omega^{o^{(\nu)}}_{1_i} - 1][(K^{wo}_2)^{(\nu)}_i \frac{\partial}{\partial P_{ow_{i+1}}} (\rho^w\varepsilon D^w S^{(\nu)}_w)_{i+\frac{1}{2}}$$

F_{37} (continued)

$$+ (K_2^{wo}K_2^{gw})_i^{(\nu)} \frac{\partial}{\partial P_{ow_{i+1}}} (\rho^g \varepsilon D^g S_g^{(\nu)})_{i+\frac{1}{2}}] + A_{37}$$

$$F_{38} = \frac{\rho_i^o}{(\Delta x)_+^2} \frac{\partial}{\partial P_{wg_{i+1}}} (\omega_2^o \tau_o)_{i+\frac{1}{2}}^{(\nu)} [P_{og_i}^{(\nu)} - P_{og_{i+1}}^{(\nu)}]$$

$$+ \frac{\rho_i^w}{(\Delta x)_+^2} \frac{\partial}{\partial P_{wg_{i+1}}} (\omega_2^o \tau_w)_{i+\frac{1}{2}}^{(\nu)} [P_{og_i}^{(\nu)} - P_{wg_{i+1}}^{(\nu)}]$$

$$+ \frac{1}{4(\Delta x)^2} \rho^o \beta_{o_i}^P (\omega_2^o \tau_o)_i^{(\nu)} [P_{og_{i-1}}^{(\nu)} - P_{wg_{i+1}}^{(\nu)}]$$

$$+ \frac{1}{(\Delta x)_+^2} \frac{\partial}{\partial P_{wg_{i+1}}} (\rho^o \varepsilon D^o S_o^{(\nu)})_{i+\frac{1}{2}} [\omega_{1_{i+1}}^{o^{(\nu)}} - \omega_{1_i}^{o^{(\nu)}}]$$

$$- \frac{1}{2\Delta x} [\omega_{1_{i+1}}^{o^{(\nu)}} - 1][\rho^o \gamma_{o_i} \frac{\partial}{\partial P_{wg_{i+1}}} (\tau_o)_{i+1}^{(\nu)} + \rho^w \gamma_{w_i} \frac{\partial}{\partial P_{wg_{i+1}}} (K_2^{wo} \tau_o)_{i+1}^{(\nu)}]$$

$$+ \frac{1}{(\Delta x)_+^2} [\omega_{1_{i+1}}^{o^{(\nu)}} - 1][\frac{\partial}{\partial P_{wg_{i+1}}} K_{2_{i+1}}^{wo} (\rho^w \varepsilon D^w S_w^{(\nu)})_{i+\frac{1}{2}})$$

$$+ \frac{\partial}{\partial P_{wg_{i+1}}} (K_2^{wo} K_{2_{i+1}}^{gw} (\rho^{g^{(\nu)}} \varepsilon D^g S_g^{(\nu)})_{i+\frac{1}{2}})]$$

$$+ \frac{1}{(\Delta x)_+^2} [\omega_{1_i}^o - 1][(K_2^{wo})_i^{(\nu)} \frac{\partial}{\partial P_{wg_{i+1}}} (\rho^w \varepsilon D^w S_w^{(\nu)})_{i+\frac{1}{2}}$$

$$+ (K_2^{wo}K_2^{gw})_i^{(\nu)} \frac{\partial}{\partial P_{wg_{i+1}}} (\rho^{g^{(\nu)}} \varepsilon D^g S_g^{(\nu)})_{i+\frac{1}{2}}] + A_{38}$$

$$F_{39} = \frac{\rho_i^o}{(\Delta x)_+^2} \frac{\partial}{\partial \omega_{1_{i+1}}^o} (\omega_2^o \tau_o)_{i+\frac{1}{2}}^{(\nu)} [P_{og_i}^{(\nu)} - P_{og_{i+1}}^{(\nu)}]$$

$$+ \frac{\rho^w}{(\Delta x)_+^2} \frac{\partial}{\partial \omega_{1_{i+1}}^o} (\omega_2^w \tau_w)_{i+\frac{1}{2}}^{(\nu)} [P_{wg_i}^{(\nu)} - P_{wg_{i+1}}^{(\nu)}]$$

$$+ \frac{1}{4(\Delta x)^2} \rho^o \beta_{o_i}^1 (\omega_2^o \tau_o)_i^{(\nu)} [P_{og_{i-1}}^{(\nu)} - P_{og_{i+1}}^{(\nu)}]$$

$$- \frac{1}{2\Delta x} [\omega_{1_{i+1}}^{o^{(\nu)}} - 1][\rho^w \gamma_{w_i} \frac{\partial}{\partial \omega_{1_{i+1}}^o} (K_2^{wo} \tau_w)_{i+1}^{(\nu)}]$$

$$+ \frac{1}{(\Delta x)_+^2} [\omega_{1_{i+1}}^{o^{(\nu)}} - 1][(\rho^w \varepsilon D^w S_w^{(\nu)})_{i+\frac{1}{2}} \frac{\partial}{\partial \omega_{1_{i+1}}^o} (K_2^{wo})_{i+1}^{(\nu)}$$

$$+ \frac{\partial}{\partial \omega_{1_{i+1}}^o} ((\rho^{g^{(\nu)}} \varepsilon D^g S_g^{(\nu)})_{i+\frac{1}{2}} (K_2^{wo} K_2^{gw})_{i+1}^{(\nu)})]$$

$$- \frac{1}{(\Delta x)_+^2} [\omega_{1_i}^{o^{(\nu)}} - 1](K_2^{wo} K_2^{gw})_i^{(\nu)} \frac{\partial}{\partial \omega_{1_{i+1}}^o} (\rho^{g^{(\nu)}} \varepsilon D^g S_g^{(\nu)})_{i+\frac{1}{2}} + A_{39}$$

Here $\Delta P_{ow_i} \equiv P_{ow_i}^{(\nu)} - P_{ow_i}^n$ and $\Delta P_{wg_i} \equiv P_{wg_i}^{(\nu)} - P_{wg_i}^n$. All parameters without superscripts are evaluated at time level n.

Lecture Notes in Engineering

Edited by C.A. Brebbia and S.A. Orszag

Vol. 1: J.C.F. Telles,
The Boundary Element Method
Applied to Inelastic Problems
IX, 243 pages. 1983

Vol. 2: Bernard Amadei,
Rock Anisotropy and
the Theory of Stress Measurements
XVIII, 479 pages. 1983

Vol. 3: Computational Aspects of
Penetration Mechanics
Proceedings of the Army Research
Office Workshop on Computational
Aspects of Penetration Mechanics
held at the Ballistic Research Laboratory
at Aberdeen Proving Ground, Maryland,
27–29 April, 1982
Edited by J. Chandra and J.E. Flaherty
VII, 221 pages. 1983

Vol. 4: W.S. Venturini
Boundary Element Method in Geomechanics
VIII, 246 pages. 1983

Vol. 5: Madassar Manzoor
Heat Flow Through Extended
Surface Heat Exchangers
VII, 286 pages. 1984

Vol. 6: Myron B. Allen III
Collocation Techniques for Modeling
Compositional Flows in Oil Reservoirs
VI, 210 pages. 1984.

Vol. 7: Derek B. Ingham,
Mark A. Kelmanson
Boundary Integral Equation
Analyses of Singular, Potential,
and Biharmonic Problems
IV, 173 pages. 1984.

Vol. 8: Linda M. Abriola
Multiphase Migration of Organic
Compounds in a Porous Medium
A Mathematical Model
VIII, 232 pages. 1984.